Programming the
Intel Edison

About the Author

Donald Norris has a degree in electrical engineering and an MBA specializing in production management. He is currently teaching undergrad and grad courses in the IT subject area at Southern New Hampshire University. He has also created and taught several robotics courses there. He has over 30 years of teaching experience as an adjunct professor at a variety of colleges and universities.

Mr. Norris retired from civilian government service with the U.S. Navy, where he specialized in acoustics related to nuclear submarines and associated advanced digital signal processing. Since then, he has spent more than 20 years as a professional software developer using C, C#, C + +, Python, Node.js, and Java, as well as 5 years as a certified IT security consultant.

Mr. Norris started a consultancy, Norris Embedded Software Solutions (dba NESS LLC), which specializes in developing application solutions using microprocessors and microcontrollers. He likes to think of himself as a perpetual hobbyist and geek and is always trying out new approaches and out-of-the-box experiments. He is a licensed private pilot, photography buff, amateur radio operator, avid runner, and, last but very important, a grandfather to a brand new baby girl—here's to you, Evangeline.

Programming the Intel Edison

Getting Started with Processing and Python

Donald Norris

New York   Chicago   San Francisco
Athens   London   Madrid   Mexico City
Milan   New Delhi   Singapore   Sydney   Toronto

Cataloging-in-Publication Data is on file with the Library of Congress

McGraw-Hill Education books are available at special quantity discounts to use as premiums and sales promotions, or for use in corporate training programs. To contact a representative, please visit the Contact Us pages at www.mhprofessional.com.

Programming the Intel Edison: Getting Started with Processing and Python

1 2 3 4 5 6 7 8 9 0   DOC DOC   1 0 9 8 7 6 5

ISBN   978-1-25-958833-4
MHID   1-25-958833-5

Sponsoring Editor	Project Manager	Art Director, Cover
Michael McCabe	Nancy Dimitry	Jeff Weeks
Editorial Supervisor	Copy Editor	Indexer
Donna M. Martone	Nancy Dimitry	WordCo Indexing Services
Production Supervisor	Proofreader	
Pamela A. Pelton	Donald Dimitry	
Acquisitions Coordinator	Composition	
Amy Stonebraker	Gabriella Kadar	

This book is dedicated to Linda Norris, who is a kind, loving, and generous person, and mother to Shauna, Heath, and Derek. She is also Mimi to grandchildren Hudson and Evangeline.

CONTENTS AT A GLANCE

CONTENTS

PREFACE

This book will serve both as an introduction to the Intel Edison computing module and also as a reliable and concise Getting Started Guide for interested readers. This computing module was introduced at the Intel Developers Forum 2014 held in San Francisco on September 10, 2014. Intel described the Edison's value as follows:

> The Intel® Edison development platform is designed to lower the barriers to entry for a range of inventors, entrepreneurs, and consumer product designers to rapidly prototype and produce IoT and wearable computing products.

The Edison's form factor, which will be described in detail later, is most definitely slated for applications demanding extremely compact hardware and, simultaneously, consuming miniscule power.

The Edison computing module is the latest in a progression of embedded technology devices that Intel has created over a long time frame. The Galileo Gen 2 development board was the most recent technology platform that just preceded the Edison. In many ways, the Galileo and Edison are quite similar except for one key aspect: The Galileo board may be used "as is," meaning that all it needs is a power supply and interconnectivity to be accessed and operated. The Edison, on the other hand, requires some type of support board to provide both power and interconnectivity. The Edison's need for a support board is the reason that I believe Intel labeled it as a computing module instead of a development board.

The Edison contains some remarkable hardware despite its very small size. It was purposefully designed to be used as a very capable embedded control module operating within an encompassing system. Intel's design philosophy was to make the module extremely compact with ultra-low power consumption. These attributes make it ideal to function as a "wearable" computer, which is described in much greater detail later in the book.

The foregoing was just a brief glimpse into what I will discuss in much greater detail in this book. Let's now delve into the Edison and see what makes it "tick."

1

Introduction

In this chapter, I will show you what makes up the Intel Edison computing module and introduce two supporting development boards that will be used in programming the Edison as well as allowing it to connect with other system components.

The Edison Computing Module

Figure 1-1 is a top view of the Edison module shown next to a U.S. nickel coin for a size comparison. It is quite small, barely larger than a typical U.S. postage stamp, with overall approximate dimensions of 34.9 × 25.4 × 3.2 mm. Under the metal cover is an Intel dual-core Silvermont Atom processor running at a 500-MHz clock speed. There is also a 100-MHz clocked Quark coprocessor included, which is designed to assist the Atom processor with input/output (I/O) operations. Unfortunately, as of the time of this writing, Intel has not released any software that will support the Quark coprocessor; therefore, it will not be discussed any further in this book. I would suggest periodically checking the Intel Edison website, http://www.intel.com/edison to see if the Quark supporting software has become available. I am sure that informative examples will also be provided to help you utilize the coprocessor.

There is also 4 GB of flash memory and 1 GB of RAM available to support the internal Edison processors. The flash memory comes preprogrammed with a Linux distribution created by Intel engineers using the Yocto framework. I will discuss this default Linux distribution in Chapter 2, in which I show you how to initially operate and communicate with the module.

There is also a Broadcom BCM43340 chip contained in the module, which implements b/g/n (11 Mbit/s, 56 Mbit/s, 100 Mbit/s internet speeds) and direct

Figure 1-1   *Top view of the Edison computing module.*

WiFi, as well as Bluetooth Low Energy (BLE) wireless communication. Both the WiFi and Bluetooth (BT) connections share the same onboard PCB chip antenna, which is visible at the lower left-hand corner in Figure 1-1. An external antenna connector using a μFL standard format is located just above the chip antenna and should be used if extended-range radio frequency (RF) operations are required. The internal chip antenna is fairly limited and will likely operate reliably only within 10 meters (m) of the WiFi access point, which is typically the wireless router in most home networks. Of course, BT communications was always designed to be close range, or not to exceed 10 m. One more point that you should know is that the antenna (internal or external) is multiplexed, or shared, between WiFi and BT operations. This might become problematic if maximum data bandwidth operations are attempted using both modes simultaneously.

The Broadcom chip also supports a hardware WiFi access-point (AP) mode, which might be very useful in certain applications. The only provision is that the module software must also support this type of operation. Fortunately, the default Linux distribution supports the AP mode, which allows for significant flexibility in configuring a network containing the Edison. Intel also provided support for BlueZ 5.0, which implements all the important and widely used BT profiles.

Now it is time to flip the module over and discuss the other side. Figure 1-2 shows the Edison's backside, where you can see another metal cover and a high-density connector.

I have already discussed what's under the cover and will now focus on the connector. It is a 70-pin connector manufactured by the Hirose company. It is considered high density because of the very tight spacing between the connector pins, which are 35 pins spread across 14 mm with 0.4 mm between pins. To put this in a common perspective, most hobbyist's solderless breadboards have a 0.1-inch, or 2.54-mm, spacing between insertion points. The contacts on the Hirose connector are about six times closer than those on a breadboard. The practical meaning for this situation is that the Edison can be used only with a development board with the matching male connector already installed on a PCB. It is just not feasible to manually solder 70 wires to a freestanding male Hirose 70-pin connector. It might be possible to solder a few wires to such a connector, using a magnifying lens and an extremely sharp-pointed soldering iron, but I think it is beyond my skill level as well as that of most of my readers. Another point worth mentioning is that the Hirose connector was not designed to be inserted and removed frequently. You can do these operations a few times, but be very careful as it is easy to damage the connecting pins by misaligning them and/or using excessive force. I believe this will not be an issue for most readers, as they likely will just mount the Edison on an appropriate development board and simply use the board with their projects.

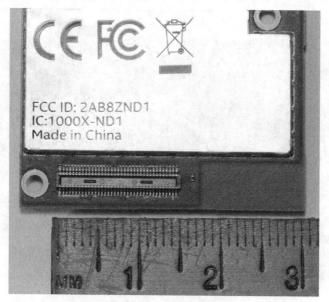

Figure 1-2 *Bottom view of the Edison computing module.*

One very nice feature of the Edison, especially as compared to somewhat similar boards such as the Raspberry Pi, is that the Edison has 40 general-purpose input/output (GPIO) pins that are available in the Hirose connector, in addition to the dedicated pins used for power and communications. I will discuss the specific pin allocations in Chapter 3. Next, I will show you the Intel Arduino Development Board, which will be the first of several support boards discussed in this chapter.

Intel Arduino Development Board

An Arduino-compatible development board was designed by Intel to allow new users to quickly use the Edison module by taking advantage of the widely known and familiar Arduino integrated development environment (IDE). This development board uses the Edison module to replace the Atmel Atmega chip used in the "normal" Arduino board. A top view of the Intel Arduino Development Board is shown in Figure 1-3 with the Edison already mounted on the board.

This appears to be a fairly complex board, but looks can be deceiving. Most of the circuitry on the board is devoted to voltage-level shifters. It turns out that the Edison uses a core voltage of 1.8 V, while the typical Arduino development package uses both 3.3 V and 5 V. Therefore, voltage level shifters are required to make

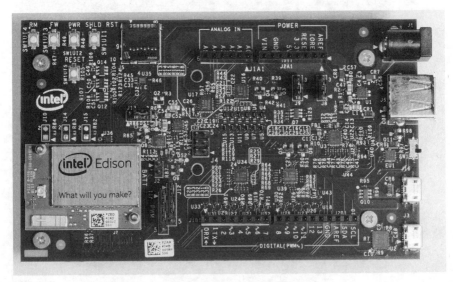

Figure 1-3   *Top view of the Intel Edison Arduino Development Board.*

the Edison function properly and safely with the much higher voltages used in normal Arduino projects. This constraint must always be kept in mind, as applying 3.3 V or 5 V directly to any of the Edison pins will instantly destroy the module. The pins are simply not protected against any inadvertent overvoltage, no matter how brief.

You should also notice that the board has sockets in place that support regular Arduino shields. Supposedly, this board will operate normal Arduino-compatible shields according to Intel marketing claims. However, it has been repeatedly reported on Edison development forums that certain shields do not function well, if at all, with this board. There are some good reasons for this situation, which I will discuss in a later chapter. For now, I would strongly suggest that you simply use a normal Arduino development board if you are looking to use a shield and avoid using this board for that particular operation. This board is not a one-for-one replacement for an Arduino development board. It was never intended to be such a board, but its purpose is still very beneficial, as it will allow you to get the Edison up and running very quickly and without much effort.

Figure 1-4 is a close-up photo of the board's surface-mounted Hirose plug that plugs into the Edison. I included this photo to reinforce my discussion on the need to use a commercially prepared mounting system and to forgo any thought

Figure 1-4   *Intel Edison Arduino Development Board's Hirose mounting plug.*

of creating your own. You can also see in this figure the two threaded posts that are used to hold the Edison in place. The threads on these posts are incredibly fine, so I strongly suggest that you do not lose the mounting nuts that came with the Edison module. I suspect you will not find any matching nuts in any local home-improvement store.

I will discuss the various board connectors in Chapter 2, in which I show you how to get this board up and running. The next development board is really bare bones, but you will still be able to use it with the Edison, albeit not as easily as with the Intel Arduino Development Board.

Intel Edison Breakout Board

As mentioned earlier, this is a no-frills board that was designed by Intel to power on the Edison using the standard USB power pins, and to provide communication with the Edison using standard USB. Figure 1-5 is a top view of this board without an Edison mounted on it.

There are four surface-mounted chips visible in the figure, two of which are dedicated to USB communication and the other two to USB voltage-level conversion between the standard 5-V USB signal levels and the Edison's 1.8-V input/output levels. You can also see four rows of 14 plated-thru holes, which may be used to connect directly to the board-mounted Hirose connector. These PCB solder points allow you to connect directly to any of the Edison module pins. Remember, that no more than 1.8 V is allowed as an input. Exceeding that level

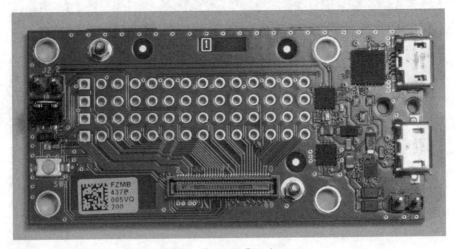

Figure 1-5   *Top view of the Intel Edison Breakout Board.*

will destroy the module. At this stage, with your limited exposure to the Edison, I would highly recommend that you avoid using any of these breakout pins.

Programming the Edison using this board will be deferred until a later chapter, as it involves using a direct Linux terminal application. I want to provide some additional background information before attempting to communicate with the Edison using this board.

Now, on to discussing the final Edison development support board in this chapter.

Sparkfun Block for Intel Edison–Console

Sparkfun (www.sparkfun.com) is a U.S. supplier specializing in open-source components and modules. At the time of Intel's Edison product announcement, Sparkfun also announced that they would make available a series of boards that would support the Edison, including the following console board.

The Sparkfun Block for Intel Edison–Console DEV-13039 board is even more no frills than the Intel Edison Breakout Board. Figure 1-6 shows a top view of this

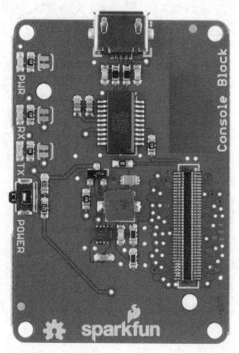

Figure 1-6   *Top view of the Sparkfun Block for Intel Edison-Console DEV-13039.*

board, where you can see it provides only USB communications and power through the USB socket.

There are no PCB connection points available as was the case for the Intel Edison Breakout Board. But there is something quite significant mounted on the bottom of the Sparkfun board. It is another male Hirose connector that allows for this console board to be connected to another Sparkfun board, providing additional functionality for the Edison. Figure 1-7 is a bottom view of the console board, which shows this mating connector.

Sparkfun has announced the following support boards, which are listed in Table 1-1 and are either available or slated to become available in the first or second quarter of 2015.

Figure 1-8 shows an example of several Sparkfun boards in a "stackup." Such a stack can offer many different levels of functionality, depending upon which boards are included in the stack.

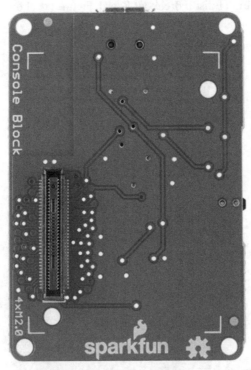

Figure 1-7   *Bottom view of the Sparkfun Block for Intel Edison-Console DEV-13039.*

Name	Model #	Description
9DOF	13033	9 degrees of freedom sensor
I2C	13034	Edison I2C bus break-out
OLED	13035	64 × 48 OLED display
Arduino	13036	ATmega 328P coprocessor
Battery	13037	Single LiPo cell and charger
GPIO	13038	General purpose i/o pin break-out
Console	13039	USB comm and power
UART	13040	RS232 with level shifter
microSD	13041	SD card holder
PWM	13042	8-chan pulse width modulation
Dual H-Bridge	13043	Drive two DC motors (1 A max)
Base	13045	Power, USB
ADC	13046	12-bit analog-to-digital converter

Table 1-1   *Sparkfun Edison Support Boards*

Later in this book I will demonstrate a simple board stack powered by the Battery board in a "wearable" project. This project will also provide a good example of the Edison's low power consumption, which allows for a cool operating system without the possibility of causing inadvertent heat injury to the person wearing the project.

Figure 1-8   *Example of a Sparkfun Edison board stack.*

Summary

I began the chapter with a brief description of what constitutes an Intel Edison computing module. The two internal processors were discussed as well as the impressive wireless communications modules, which provide WiFi and Bluetooth, simultaneously. The Hirose connector was also studied in detail, as that is a key component in how the Edison connects with any external development board or module.

Three development boards were next described, starting with the Intel Arduino Development Board. I mentioned that the Intel Arduino board was probably the fastest and easiest way to start programming and using the Edison, especially for new users. This board uses the well-known and easy to master Arduino open-source integrated development environment.

The next board described was the Intel Edison Breakout Board, which really is quite minimalist as compared to the Arduino board. This board provides only power and USB communication between the Edison and a Linux-based terminal program running on an external laptop. This operational mode will be discussed in a later chapter.

Finally, I showed you one of the stackable Sparkfun boards that provides only USB and power to the Edison as did the previous board. However, you can combine several Sparkfun boards in a "so called" stack to provide many different functions to support a variety of projects.

The next chapter will explore how to set up and use the Edison in an Arduino development environment.

2

Getting Started with the Intel Edison Arduino Board

In this chapter, I will show you how to install and configure the Intel Edison Arduino integrated development environment (IDE) so that you can connect with and program the Edison module that is connected to the Intel Edison Arduino Development Board. I will also discuss two ways to power the development board as well as two approaches for USB communication associated with this board.

A simple LED-blinking example will also be demonstrated, which will prove that the IDE, the Edison, and the development board all function together as expected.

Intel Edison Arduino IDE

An Arduino IDE was created by Intel and specifically designed to operate with their Edison Arduino Development Board. This IDE is an emulation of the regular, open-source Arduino IDE that is available from the http://arduino.cc website. The Intel version can be downloaded by going to http://www.intel.com/edison and navigating to the Downloads & Documentation page. There are four versions of the IDE available in the Downloads page, and all were at revision 1.5.3 at the time of this writing. You should select the appropriate version that matches your host computer OS that you will use with the development board. I chose the Mac OS X version, as that matched my host system. There are 32- and 64-bit Linux versions available as well as a Windows version. Note that all the versions

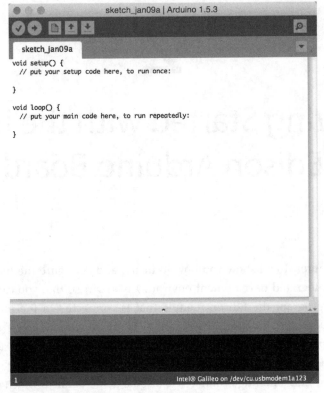

Figure 2-1 *Opening screenshot for the Intel Arduino IDE.*

are compressed and will have to be extracted before the actual software can be installed on the host.

Figure 2-1 is an opening screenshot of the Intel Arduino IDE running on my MacBook Pro. You should notice that the IDE automatically creates a blank sketch template with a name containing the current day of the month, which, in this example, is sketch_jan09a. Also, note the IDE revision number after the Arduino name in the title bar. A regular Arduino sketch would display a revision of 1.0.5, which was current at the time of this writing. Be particularly mindful of the revision shown, especially if you have installed both the regular and the Intel IDEs on your host. The regular IDE will not connect with or program the Intel Arduino Development Board, and the reverse is true for the Intel IDE attempting to connect with an Arduino board. I will also show you how to work with a blank sketch to create a program, but first I need to demonstrate how the development board can be powered.

Powering the Arduino Development Board

Figure 2-2 shows the power and USB connectors mounted on the edge of the development board.

There are two ways to power the development board:

1. Use the 2.1 mm barrel jack with a 7.5-V to 12-V external DC power supply.

2. Connect a powered USB cable to the OTG type B micro USB connector labeled J16 and located just to the left of the edge-mounted slide switch.

Connecting an external power supply to the barrel jack is the preferred way to supply power, especially if you plan on using wireless communications and/or if your project will need to supply a substantial amount of current through the GPIO pins. Providing power using a USB cable technically limits the board to a

Figure 2-2 *Edge view of the development board.*

maximum current of 500 ma, which is the USB standard. The board itself will take about 200 ma when operating wireless communications, which leaves a maximum of 300 ma for all other requirements. Operating a single LED using a GPIO pin will typically consume anywhere from 20 to 30 ma, so you can see your current requirements rapidly accumulate. Straining a power supply can result in some very strange and odd development board behavior, so it is always wise to ensure that you have a stable and more than adequate power source.

NOTE   *It will not harm the board if you use the barrel connector for the main power supply and, at the same time, connect a powered USB cable to J16.*

USB serial communications will next be discussed, as that is also a key element in setting up the Arduino development environment.

USB Communications

Figure 2-3 is a block diagram of the Intel Arduino Development Board in which you can see several USB ports diagrammed at the bottom, center of the figure.

You will need to use the J16 micro USB connector as the link between the development board and the host computer. Notice that the line on the block diagram going between the USB Mux block and the Edison block is labeled as USB OTG, where OTG is an acronym for On-The-Go, a USB specification issued in 2001. USB OTG allows USB devices, such as digital audio players or similar mobile devices, to act as a host, which allows other USB devices, such as thumb drives, to function with them.

The Edison implements the OTG specification, which allows it to both read from and write to a mass storage device, such as a thumb drive, as mentioned earlier. The Edison may also appear as a mass storage device when appropriately configured and connected to a host computer. In other words, the USB OTG specification allows a device to function either as a master or slave, depending upon the application. In the OTG configuration, the device controlling the USB communications link is referred to as the master, and the device being controlled, as the slave.

The OTG specification is the reason that the development board has two of the three USB connectors visible in Figure 2-2. The USB connector to the immediate right of the slide switch is a standard "A" female USB connector and will

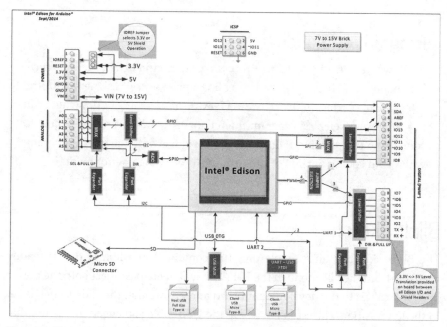

Figure 2-3 *Intel Arduino Development Board block diagram.*

be the connector used if the board is in the master or host configuration. Conversely, the micro "B" female connector to the immediate left of the slide switch will be the connector used if the board is in the slave or device configuration. Naturally, you must slide the switch to the appropriate side and use the proper connector to enable the desired OTG configuration.

Given all of the above background discussion, it is really important to know only that you must connect a micro-to-standard USB cable from J16 to the host computer with the slide switch pushed to the left

The remaining micro USB connector located on the far left of the board, as shown in Figure 2-2, is a USB UART connected to a serial FTDI chip, a configuration that would be used for a dedicated client-type application. The slide switch has no impact on this connector because it is always enabled (or not) through the Edison software.

It is time to demonstrate a simple Arduino program now that I have explained both the power and the USB communications aspects of the development board.

Program or Sketch

Code written in the Arduino/Processing environment is called a *sketch*. The name was chosen by the Processing open-source designers with the intention of playing off the artistic endeavors of sketching or drawing. I will use the term sketch or program interchangeably while discussing how to code using the Arduino IDE. However, I will use only the word, program, when I am coding outside of the Arduino/Processing environment.

Blink Sketch

The Blink sketch is an example program contained in a library that is readily available from the File menu in the IDE. You will need to connect both the power and USB cables to the development board and ensure that the slide switch is set to the left position, closest to J16 where the USB cable is plugged in. Ensure that the USB cable is plugged into the host computer. Next, follow these steps to load and run the Blink sketch:

1. Start the IDE application on the host computer, and you should see the display as shown in Figure 2-1.

2. Select the Edison board by clicking
    Tools -> Board -> Edison
    (**Note**   *You will get perplexing results by attempting to run the Blink sketch with the Galileo board selected.*)

3. Select the appropriate serial port for your host computer. In my case, it was /dev/cu.usbmodem1d113 for the MacBook Pro. Click on Tools -> Serial Port -> <your appropriate serial port>

4. Load the Blink sketch by clicking
    File -> Examples -> 01.Basics -> Blink

You should now see the Blink sketch code displayed on the host, as shown in Figure 2-4.

Loading and executing the sketch is very easy; all you need to do is click on the right-facing arrow located in the menu bar section of the IDE window. An

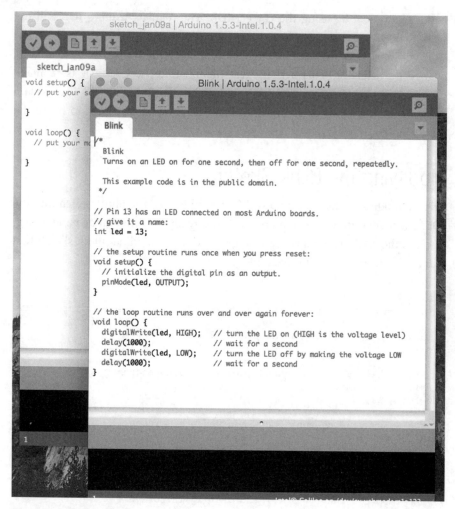

Figure 2-4   *Blink sketch.*

LED should start blinking on/off every second after a brief time delay during which the sketch is compiled and transferred to the development board. The blinking LED is logically referred to as D13 and is located on the board about 0.5 inches to the right of the top edge of the Edison module. Recheck the IDE settings if you do not see a blinking LED. The most likely cause is either you did not select the Edison board or you have an incorrect serial port selected. Of course, you must be using the correct IDE. The regular Arduino IDE will not work with this development board, as I mentioned above.

It will next be a worthwhile exercise to modify the Blink sketch to slightly change its behavior. This activity will demonstrate how easy it is to make program changes and test them on the development board. I will also explain some of the key features of this sketch, while showing you how to make the modifications.

Modifying the Blink Sketch

Prior to modifying the code, you will first need to load the Blink sketch into the IDE. All changes to the sketch are entered directly into the Editor pane, which displays the code. The following listing is a copy of the initial Blink sketch that was loaded from the default Examples library.

```
/*
  Blink
  Turns on an LED for one second, then off for one second, repeatedly.

  This example code is in the public domain.
*/

// Pin 13 has an LED connected on most Arduino boards.
// give it a name:
int led = 13;

// the setup routine runs once when you press reset:
void setup() {
  // initialize the digital pin as an output.
  pinMode(led, OUTPUT);
}

// the loop routine runs over and over again forever:
void loop() {
  digitalWrite(led, HIGH);   // turn the LED on (HIGH is the voltage
                             // level)
  delay(1000);               // wait for a second
  digitalWrite(led, LOW);    // turn the LED off by making the voltage
                             // LOW
  delay(1000);               // wait for a second
}
```

At the start of the listing are some descriptive words regarding what the sketch was designed to do. These are called comments and are differentiated from normal code in one of two ways.

1. Multiline comments are contained within the symbols /* ... */ as shown in this example:

    ```
    /* some text
    more text
    last of the text */
    ```

2. A single line comment just follows these symbols //, as shown in this example:

    ```
    // some text
    ```

Either single line comments are on their own line in the code, or they always follow any actual code and are terminated by a carriage return (CR) and line feed (LF).

It is always a good idea to include comments in your programs or sketches. It helps refresh your memory when you return to the program after some significant time lapse, and it also is very helpful for someone new to your program in interpreting what you wanted to accomplish with it.

There are two "parts" to this sketch that are common to all sketches. They are the setup and loop methods and are shown in the code as follows:

```
void setup() {
...
}
```

and

```
void loop() {
...
}
```

The ellipses contained within the braces are simply placeholders representing the actual code. These two methods are also known as functions and may be thought of as code that collectively does something in support of the overall sketch behavior. The setup method as the name implies puts in place certain preconditions that are necessary for the sketch to accomplish its stated purpose. In this case, the setup method changes GPIO pin 13 from its default input mode to an output mode so that it can control an LED that is permanently connected to it on the development board. The setup method is called or activated only one time, as that is all that's needed for this configuration.

The next method is named `loop`, and as its name implies, is repeatedly called or activated for as long as the sketch runs. The code contained within the braces does four things as I discuss below:

1. `digitalWrite(led, HIGH);`   `// turns LED on (HIGH is the voltage level of 3.3 V)`

2. `delay(1000);`   `// wait for a second (number represents millisecs)`

3. `digitalWrite(led, LOW);`   `// turns LED off (LOW is the voltage level of 0 V)`

4. `delay(1000);`   `// wait another second`

This method repeatedly blinks the LED on and off for a one second duration in each state. The Blink sketch is automatically compiled, downloaded to the development board, and executed by the underlying Arduino operating code that is activated when you click on the right-facing arrow in the IDE.

Let's now modify the code so that the LED blink rate is twice as fast. Based upon my previous discussion, I believe you can see that this change can easily be accomplished by reducing the time delays from one second to one-half second for both the LED on and off times. In other words, changing both delay statements to:

```
delay(500);
```

I made these changes in the editor and then clicked on the arrow to observe the new behavior. I subsequently observed the LED blinking twice as fast after a few seconds delay while the sketch was recompiled and downloaded into the development board.

When you try to close the IDE, you will be prompted to save the modified sketch, as the IDE recognizes that changes were added. In this case, I elected not to save the modified sketch, as the changes were minor and could easily be added the next time I ran the Blink sketch. However, you will likely have a situation in the future in which you do make extensive modifications to a sketch and you should save it with a new name indicative of what the modified sketch accomplishes. For this previous example, I could envision changing the sketch name to FastBlink.

Summary

This now concludes what I wanted to demonstrate in this chapter regarding how to set up and operate the Intel Arduino Development Board with the Intel version of the Arduino IDE. The next chapter goes into much further detail regarding how to program the Edison with the Arduino IDE and how to create your own programs or sketches.

3

Working with Processing and the Intel Arduino IDE

In this chapter, I will cover some of the basic concepts that are important in creating sketches that can operate the Edison, using both the Intel Edison Arduino IDE and the Intel Arduino Development Board. I will also incorporate specifics related only to the Edison hardware, which should help focus and differentiate this chapter's context from the myriad of other Arduino how-to-do-it books that are in the marketplace.

The Processing Language and the Intel Edison Arduino IDE

Processing was the language created by the Arduino development team to program the original open-source Arduino hardware. It is based primarily on the C language, which has been in existence for many years and is still quite relevant for current embedded development projects. Processing also includes some object-oriented components, which makes it somewhat similar to the C++ language, but far less complex and comprehensive. The Intel software development team created an emulation of the original Arduino IDE so that it could program and operate the Edison module when it was attached to the Intel Edison Arduino Development Board, which I will, from now on, simply refer to as the *dev board* to save a lot of text entry.

The Intel and Arduino IDEs are different in their underlying make-up as I mentioned in the previous chapter. The most important concept, which you should know, is that the Intel IDE creates a program that runs in the Edison's Linux operating system and tries to function as a real hardware version of the Arduino. This concept is known as emulation and is not, and never can be, the same as real hardware. One of the consequences of emulation is that access to GPIO pins for reading and writing in their Processing implementation is much different from the original IDE Processing language implementation.

Again, as stated in the previous chapter, you cannot use the original Arduino IDE to program and run the Edison mounted on a dev board, nor can you use the Intel Edison IDE to program or run an Arduino board. The Intel Edison IDE along with its Processing implementation was really created to allow for rapid application development with the Edison by using a well-known and easily understood programming environment. Another programming approach using the built-in Python language will be discussed later in this book. That approach is much faster and "cleaner," as it does not depend upon an emulator to function.

The good news is that the Processing language is identical for both implementations when you program at a higher abstract level. This means the program logic can be created more or less independently of whether you are actually using a dev board or an Arduino.

Now it is time to cover some basic program development. You can skip the next sections if you are familiar with and comfortable programming the Arduino. However, there might be a tidbit or two in the following sections that will refresh your Arduino programming experience.

Processing Language Basics

Let's start by stating that there are only three ways that a program will execute from start to finish. These are normally called flow of control and are listed below:

1. Step-by-step
2. Conditional (Selection)
3. Repetitive (Looping)

The first, *step-by-step*, is the default way a program operates. Program execution starts at the very first statement and then goes through every subsequent statement in the order in which the program statements were entered until the last or end program statement is encountered. What happens next is totally dependent upon the operating system (if any) that is being used to control the processor. I will cover this topic further when the repetitive flow of control is discussed.

The next type of control is termed *conditional,* or *selection,* and is frequently used when a decision must be made regarding whether the execution stops being step-by-step and instead goes, or branches, to a specific location within the program. Most often, the if or if/else statements are used to implement this behavior.

The last type of control is termed *repetitive,* or *looping,* which means that an instruction or group of instructions is repeatedly executed indefinitely, or until a specific condition is met. Sometimes the condition is an external event that signals the program to stop looping. The for and while statements are most commonly used in creating a loop in Processing code. Looping may also be implemented in a method that is repeatedly called by an operating system or similar background process. The loop method, which was first discussed in Chapter 2, is contained in every Arduino template. It simply repeats indefinitely until processor execution stops, or the power is interrupted.

I will next discuss input and output statements, as these are the most common types of operations that are executed by processors such as the Edison or Arduino.

Input and Output Statements

Turning input and output lines on or off are the tours de force, or main reasons, that embedded processors exist. Without such operations, it would be impossible for the Edison to control anything. Input/output (I/O) lines are key elements in any project that you build. I have replicated the pinout diagram from the Intel Edison's User's Guide in Figure 3-1. The User's Guide is available as a PDF download from the Intel Edison website. It is an invaluable reference that you will continually go back to as you construct projects.

If you look closely at this figure, you will see 40 pins that start with the label GP. These are the general-purpose input/output (GPIO) pins that are typically programmable as either an input or output. Note, that most pins also have another descriptor that follows the pin number. This is the default behavior for that pin whenever the processor is powered on or reset. Also, not all of these GPIO pins are available on a particular development board, as it depends greatly upon what purpose the board is designed to serve. Most of the GPIO pins are

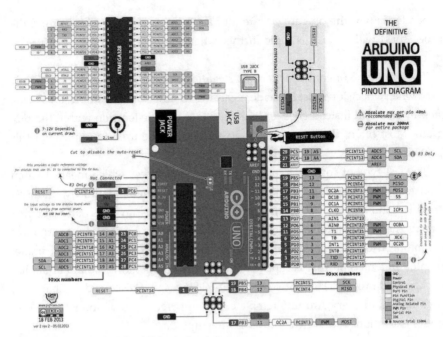

Figure 3-1 *Sparkfun's Intel Edison pinout diagram.*

available on the Arduino dev board, while none are available on the Sparkfun console board. However, remember that the console board is stackable so that all the GPIO pins, as well as the remaining 30 pins, are carried through the Hirose connectors to any other stacked boards. This means that any boards in the stack can break out the GPIO pins, as necessary, to carry out their designed function.

It is now important to clarify the dev board pin numbering prior to discussing the Processing instructions that control the user-available pins on the board. There are 20 GPIO pins available on the shield sockets, which are shown in the figure and designated as IOxx. These are the exact pin numbers that you should use to manipulate the corresponding pins when writing a Processing sketch. All the alternate functions assigned to each pin are also clearly shown in the figure.

Processing has several instructions that are used to configure and control the GPIO pins. These are listed in Table 3-1 with a brief description on how to use each instruction:

It is also critical to have memory locations to store and retrieve data. This is the purpose of data variables, which are discussed next.

Instruction	Description
pinMode(pin, I/O)	Sets the direction of a specified *pin* to either an *INPUT* or *OUTPUT*
digitalRead(pin)	Read the digital value of the specified *pin*, either a 1 or 0
digitalWrite(pin, value)	Write the value to the specified *pin*, using the constants *HIGH* or *LOW* or the numeric values of 1 or 0
analogRead(pin)	Read the analog value of the specified *pin*. The value read depends upon the analog voltage with a range of 1023 for 5-V to 0 for 0-V input. Only pin numbers 0 to 5 are valid and no pinMode() method call is required to support an analogRead() method call.

Table 3-1   *Processing I/O Instructions*

Data Variables

Non-trivial sketches cannot be written without memory locations to both store and retrieve data, whatever the format. This is the purpose that data variables serve. The Processing language is based on the C language, which is classified in computer science terms as *strongly typed*. This means that only properly formatted data may be stored in predefined locations for that formatted data, avoiding the situation where a round peg is being forced into a square hole. Table 3-2 details the Processing data types and how to declare them.

Type	Data Declaration	Remarks
Integer	int num;	Whole numbers from −32,768 to 32,767
Unsigned integer	unsigned int num;	Positive numbers from 0 to 65,535
Large integer	long num;	Whole numbers from −2,147,483,648 to 2,147,483,647
Unsigned large integer	unsigned long num;	Positive numbers from 0 to 4294967295
Real number	float num;	Real numbers from 3.4028235E+/−38 to −3.4028235E+/−38
Real number	double num;	Same as float for Processing
Logical	boolean num;	true or false (0 = false, 1 = true)
Character	char num;	Whole numbers from −128 to 127
Byte	byte num;	Whole numbers from 0 to 255

Table 3-2   *Processing Data Types*

While typical Arduino sketches do not normally require data to be displayed to the user, it is sometimes helpful to view real-time data and other information related to an executing sketch. Processing uses the string variable to store an array, or collection of characters, that may be displayed as either text or numeric information. Strings are declared and defined in several ways depending upon how they will be used in a sketch. Strings that are constant are declared as literals as follows:

```
String title = "An example sketch";
```

A string that can hold variable text is simply declared as follows:

```
String name;
```

There are various ways that a string variable can have its text or numeric data assigned. Often, text and numeric data are programmatically generated and then added one character at a time to the string. Data may also be transferred directly into the string from a keyboard. How a string is populated really depends upon what purpose the sketch string serves and the design of the user interaction. No matter how strings are created, they are displayed to the user by the Serial Monitor, which is activated by clicking on the magnifier glass icon located in the upper right-hand corner of the Arduino IDE. I will next present an example sketch that demonstrates most of the basic concepts that were just discussed, including how to use the Serial Monitor.

Average Voltage Measurement Sketch

This sketch will compute an average level for a signal waveform input. The level for a perfect sine wave signal should be 0.5 times the peak level of the sine signal, assuming that the sine wave is unipolar, i.e., there are no negative voltage levels and the downward peak is set at 0 V.

This sketch was designed to sample a sine wave signal at four times the input frequency, which should provide a good sampled representation of the input waveform. The sampling will also take place over a five-second duration, which will allow for 200 samples to be generated, given the 10-Hz input frequency.

I used a model 3406B Picoscope to both generate and observe a 10-Hz sine wave. The 3406B is a four-channel, USB-controlled oscilloscope that also contains an arbitrary function generator (AFG) that had been preset to generate a sine wave for this experiment. Figure 3-2 is a screenshot of the generated 10-Hz sine wave, which was offset by 1 volt to account for the dev board's ADC, which

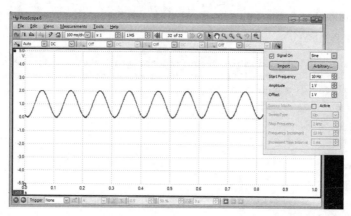

Figure 3-2   *10-Hz input sine wave.*

can handle only positive voltages between 0 and 5 V. The AFG sine wave will now be 2 V peak, which means the average is 1 V.

The signal generator output was connected to IO0 on the dev board, which is also the first analog input channel A0, as depicted by the Figure 3-3 Fritzing diagram.

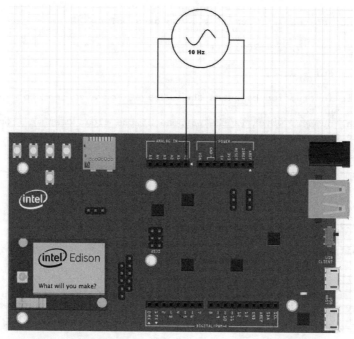

Figure 3-3   *AFG connection diagram.*

The following is the fully commented sketch listing that measured the sine wave average level.

```
/* Declare all the variables required for this sketch. Notice that an
integer array, int data[200], was declared to hold all the data used
in the average computation.
*/

int analog = 0;
long sum;
int data[200];
float ave;

void setup() {
  // Need to start the serial monitor at 9600 baud
  Serial.begin(9600);
}

void loop() {
  // First we need to collect 200 data points, all integer values up
     to 1023 = 5 V
  for (int i = 0; i < 200; i++)
  {
    data[i] = analogRead(analog); // Reads pin 0
    delay(25); // Wait about 25 millisecs between data readings
  }
  // Next, compute the sum of all the data points
  sum = 0;
  for (int i = 0; i < 200; i++)
  {
    sum = sum + data[i];
  }
  // This instruction computes the average. You also need to convert
     the sum to a
  // real number, which is the purpose of the float cast in front of
     the sum variable
  ave = ((float)sum) / 200.0;

  // This instruction converts the above number to the equivalent
     voltage
  ave = 5.0 * (ave / 1023);

  // These are two print statements for the serial monitor
  Serial.print("Average level = ");
  Serial.println(ave);

}
```

Figure 3-4 is a screenshot of the Serial Monitor taken while the sketch was running. As you can readily see from the figure, the average level was 0.99 V,

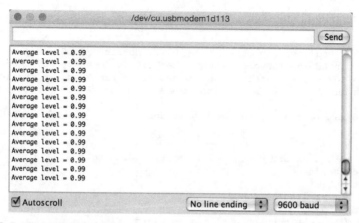

Figure 3-4 *Serial Monitor screenshot.*

which is very close to the predicted 1-V average level for the input sine waveform. The level display is updated every five seconds, which was also preset in the sketch.

The foregoing sketch demonstrated some key points with regard to creating an Arduino sketch. Variables were initially declared that held real-time data as well as the final results. An analog input was also used to sample the waveform, and an array of integers was set up to store 200 sample values that were then used to compute the average signal level. Finally, the end result was displayed using the built-in Arduino IDE Serial Monitor.

Believe it or not, I have already covered most of the instructions that you will need to read sensors and/or control motors using the dev board. I just need to demonstrate one more relatively simple sketch that reads a switch and takes an action based upon the state of a switch, i.e., pressed or not pressed.

Switch Demo Sketch

This sketch is a modification of the Blink sketch with the addition of a push-button input. When the push button is pressed, the digital input will go from a high level to a low level, and a separate LED connected to pin IO2 will then start to blink twice a second for as long as the push button is pressed. Also, the LED connected to pin 13 will not blink when the push button is pressed.

The modified Blink sketch, which I named Blink2, is listed below:

```
/*
  Blink2. Modification of Blink by D. J. Norris 1/2015
  Turns an LED on for one second, then off for one second, repeatedly.
  Polls for a push-button press connected to pin 7 and if pressed will
  blink an LED connected to pin 2 for as long as the button is pressed.

  This example code is in the public domain.
 */

// Pin 13 has an LED connected on most Arduino boards.
// give it a name.
int led = 13;
// Pin 2 must also be named.
int led2 = 2;
// Likewise, the push button is also named.
int pb = 7;

// the setup routine runs once when you press reset.
void setup() {
  // initialize the digital pin as an output.
  pinMode(led, OUTPUT);
  // initialize the second LED pin as an output.
  pinMode(led2, OUTPUT);
// NOTE: By default, all digital pins are inputs, so pin 7 does not
// require a pinMode initialization as an input.
}

// the loop routine runs over and over again forever.
void loop() {
  if(digitalRead(pb))        // call the fastBlink method when the pb
                             // is pressed
  {
    fastBlink();
  }
  else
  {
  digitalWrite(led, HIGH);  // turn the LED on (HIGH is the voltage
                            // level)
  delay(1000);              // wait for a second
  digitalWrite(led, LOW);   // turn the LED off by making the voltage
                            // LOW
  delay(1000); // wait for a second
  }
}

// a separate blink method for the second LED.
void fastBlink() {
  digitalWrite(led2, HIGH); // turn on the second LED
  delay(500);               // wait for .5 seconds
```

```
  digitalWrite(led2, LOW);      // turn off the second LED
  delay(500);                   // wait for .5 seconds
}
```

You will need to wire a push button and a second LED with supporting resistors, as shown in Figure 3-5, in order to run the Blink2 sketch. Remember that the pin-13 LED is permanently installed on the dev board.

Please note that I created a separate method named `fastBlink` to blink the LED connected to pin 2. This method is logically outside of the `loop` method; however, it will be called whenever the push button is pressed. This approach to programming results in more readable and understandable code and should always be kept in mind.

My next sketch will demonstrate how to control an actuator, which, in this example, is a mini-servo. I am showing this sketch because I will be using servos

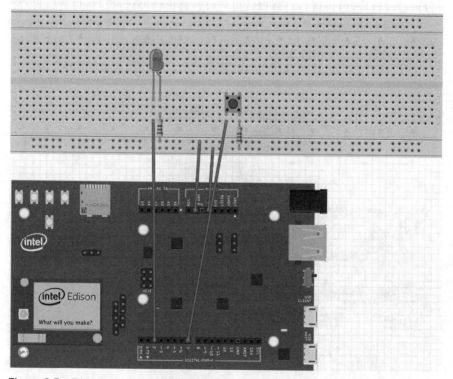

Figure 3-5   *Fritzing wiring diagram for the Blink2 sketch.*

in the Chapter 4 project and it would be useful for you to gain a little experience with this type of device.

Mini-Servo Sketch Example

A servo may be considered a specialized electric motor that usually has a restricted range of motion, but not always. Servos utilize internal circuits to control their range of motion. Some servos can rotate $+/-$ 180°, while others can continuously rotate either clockwise or counter-clockwise, just as a regular electric motor functions. The mini-servo used in this example is a limited range-of-motion type, but should suffice to demonstrate all that you will need to know regarding controlling a servo with the dev board.

Figure 3-6 is a Fritzing diagram showing the connections between the servo, dev board, and a battery supply for the servo.

Figure 3-7 shows the physical setup for the mini-servo connected to the dev board with only two leads, or wires. The servo does have an additional power lead because it requires a 6-V to 9-V power supply that has at least 200 ma current capability, which cannot be provided by the dev board.

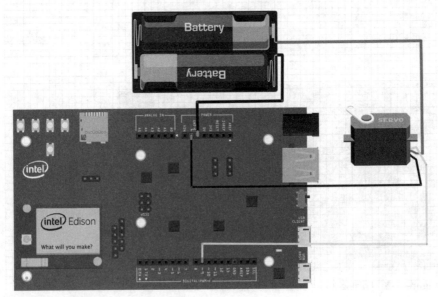

Figure 3-6 *Mini-servo connection diagram.*

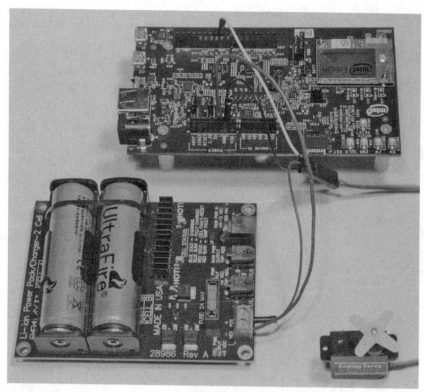

Figure 3-7   *Mini-servo connected to a dev board.*

The servo power supply shown in the figure is a two-cell LiPo battery/charger board that was purchased from Parallax, where I also purchased the robot kit featured in Chapter 4. Each LiPo cell is 3.7 V, which, when connected in series, provides 7.4 V for the mini-servo. The battery board also provides a peak current in excess of 2 A, which is way more than the mini-servo requires.

The sketch controlling the mini-Servo is named Sweep and is available from the IDE Examples library. You can load it in the same manner that you used for the Blink sketch, and you will find this sketch in the Servo sub-library. The listing below shows the Sweep sketch.

```
// Sweep
// by BARRAGAN <http://barraganstudio.com>
// This example code is in the public domain.

#include <Servo.h>
```

```
Servo myservo;        // create servo object to control a servo
                      // a maximum of eight servo objects can be created

int pos = 0;              // variable to store the servo position

void setup()
{
  myservo.attach(9);       // attaches the servo on pin 9 to the servo
                              object

}

void loop()
{
  for(pos = 0; pos < 180; pos += 1)   // goes from 0 degrees to 180
                                          degrees
  {                                   // in steps of 1 degree
    myservo.write(pos);               // tell servo to go to position
                                         in variable 'pos'
    delay(15);                        // waits 15 ms for the servo to
                                         reach the position

  }
  for(pos = 180; pos >= 1; pos -= 1)     // goes from 180 degrees to
                                            0 degrees
  {
    myservo.write(pos);                 // tell servo to go to position
                                           in variable 'pos'
    delay(15);                          // waits 15 ms for the servo to
                                           reach the position

  }
}
```

This sketch contains several new programming techniques that are worth discussing. First, notice the statement near the top of the listing:

```
#include <Servo.h>
```

The '#' symbol in front of the word include is a compiler directive that instructs the IDE to add a special file named Servo.h, which contains all the needed attributes and method names used in a companion file named Servo.cpp. The extension .cpp indicates the latter file is a C++ file, containing an object-oriented (OO) template from which a Servo object can be instantiated. It is not my intention to go into depth on how to program with OO constructs, but it is important to realize what is happening in this small sketch.

Immediately after the include directive comes the creation of a reference to a Servo object, which is named myservo. Below is the instruction that accomplishes this task:

```
Servo myservo;
```

The reference, myservo, is an object of the Servo class from which a variety of methods may be called. Calling a method in this fashion is done as follows:

```
myservo.attach(9);
```

This instruction associates pin 9 with the myservo object so that all subsequent digital commands from the Servo object are directed to that pin. There can be many methods contained within the Servo class depending upon how its responsibilities are defined. Responsibility in this sense means how a class meets requirements imposed upon it during the design phase. The Arduino website is a great reference for determining what methods or behaviors are contained within a specific class. This reference is also known as an application programming interface (API) and should be used whenever you have a question regarding how a particular Arduino class can be used or modified. I have included below a portion of the Servo class API for the attach method as an example of how useful the API is:

```
attach()
Description
Attach the Servo variable to a pin. Note that in Arduino 0016 and
earlier, the Servo library supports servos on only two pins:
9 and 10.
Syntax
servo.attach(pin)
servo.attach(pin, min, max)
Parameters
servo: a variable of type Servo
pin: the number of the pin that the servo is attached to
min (optional): the pulse width, in microseconds, corresponding to the
minimum (0-degree) angle on the servo (defaults to 544)
max (optional): the pulse width, in microseconds, corresponding to the
maximum (180-degree) angle on the servo (defaults to 2400)
Example
#include <Servo.h>
Servo myservo;
void setup()
{
  myservo.attach(9);
}
void loop() {}
```

Obviously, the above code example does not function or accomplish anything other than to logically connect the myservo object to pin 9.

The next and final sketch in this chapter is actually a failed experiment in which I tried to interface an ultrasonic ping distance sensor with the dev board.

I elected to include it in this chapter to point out a key limitation contained within the Edison Arduino emulation as compared to a real hardware Arduino board. However, I will demonstrate a working sensor setup using an Arduino Uno board as a controller.

Ping Sensor Sketch

This sketch demonstrates my attempt to interface a fairly sophisticated sensor with the dev board using the Processing language. The sensor I used is called a Ping and is sold and distributed by Parallax. Figure 3-8 shows both an oblique front view and a rear view of this sensor.

The piezoelectric transmitter and receiver are clearly visible on the front view, and a microprocessor with supporting circuitry is visible in the back view of this figure. This sensor functions essentially like a bat's natural sensory system does. The sensor emits a very brief 40 kHz ultrasonic pulse and listens for a return echo. If it detects a returning echo, it will immediately transition a timing pulse from high to low. The duration of the timing pulse is proportional to the distance between the sensor and the reflector because the timing pulse was started at the instant the original ultrasonic pulse was transmitted. An external controller will then measure the timing pulse length in units of microseconds and convert that number to a physical distance. That's easy to do given that the speed of sound in air is a constant. A block diagram illustrating this arrangement is shown in Figure 3-9.

Figure 3-8 *Ping sensor views.*

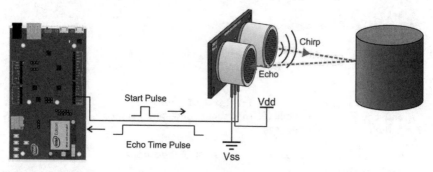

Figure 3-9 *Ping sensor operational block diagram.*

The sketch, which interfaces to and operates this sensor, is named Ping. Its listing is shown below:

```
/* Ping))) Sensor

   This sketch reads a PING))) ultrasonic rangefinder and returns the
   distance to the closest object in range. To do this, it sends a
   pulse to the sensor to initiate a reading, then listens for a pulse
   to return. The length of the returning pulse is proportional to
   the distance of the object from the sensor.

   The circuit:
    * +V connection of the PING))) attached to +5 V
    * GND connection of the PING))) attached to ground
    * SIG connection of the PING))) attached to digital pin 2

   http://www.arduino.cc/en/Tutorial/Ping

   created 3 Nov 2008
   by David A. Mellis
   modified 30 Aug 2011
   by Tom Igoe
   further modified 15 Jan 2015
   D. J. Norris

   This example code is in the public domain.
*/

// this constant won't change. It's the pin number
// of the sensor's output:
const int pingPin = 2;
```

```
void setup() {
  // initialize serial communication:
  Serial.begin(9600);
}

void loop()
{
  // establish variables for duration of the ping,
  // and the distance result in inches and centimeters:
  long duration, inches, cm;

  // The PING))) is triggered by a HIGH pulse of 2 or more
  //   microseconds.
  // Give a short LOW pulse beforehand to ensure a clean HIGH pulse:
  pinMode(pingPin, OUTPUT);
  digitalWrite(pingPin, LOW);
  delayMicroseconds(2);
  digitalWrite(pingPin, HIGH);
  delayMicroseconds(5);
  digitalWrite(pingPin, LOW);

  // The same pin is used to read the signal from the PING))): a HIGH
  // pulse whose duration is the time (in microseconds) from the
  //   sending
  // of the ping to the reception of its echo off of an object.
  pinMode(pingPin, INPUT);
  duration = pulseIn(pingPin, HIGH);

  // convert the time into a distance
  inches = microsecondsToInches(duration);
  cm = microsecondsToCentimeters(duration);

  Serial.print(inches);
  Serial.print("in, ");
  Serial.print(cm);
  Serial.print("cm");
  Serial.println();

  delay(100);
}

long microsecondsToInches(long microseconds)
{
  // According to Parallax's datasheet for the PING))), there are
  // 73.746 microseconds per inch (i.e. sound travels at 1130 feet per
  // second).  This gives the distance travelled by the ping, outbound
  // and return, so we divide by 2 to get the distance of the
  //   obstacle.
  // See: http://www.parallax.com/dl/docs/prod/acc/28015-PING-v1.3.pdf
  return microseconds / 74 / 2;
}
```

```
long microsecondsToCentimeters(long microseconds)
{
    // The speed of sound is 340 m/s or 29 microseconds per centimeter.
    // The ping travels out and back, so to find the distance of the
    // object we take half of the distance travelled.
    return microseconds / 29 / 2;

}
```

Figure 3-10 is a screenshot of the Serial Monitor with the Ping sketch loaded and running on the dev board. As you can readily see, all that is shown are zero distances, indicating that the dev board does not function with this particular sensor. This odd situation arises out of the fact that the sensor initially requires that a very short duration start pulse, on the order of several microseconds, be sent out of the signal line. This same line is then switched from an output to an input, and the Edison processor starts waiting for the signal line to transition from high to a low, all the while running a timing clock in the background. This whole process requires very rapid timing, which simply is not supported by the Intel emulation that makes the Edison appear like it is functioning as an Arduino. This is a sad fact that users must know: emulations are limited and will not precisely match the real hardware being emulated, especially in situations where fast, real-time clock timing is involved.

I next connected an Arduino Uno to the sensor and loaded it with the same sketch. Note that I had to use the regular Arduino IDE to program the Uno. The

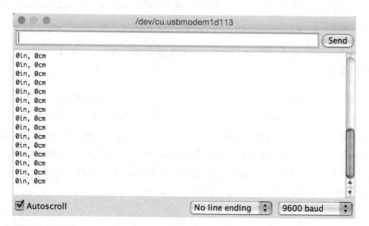

Figure 3-10   *Serial Monitor screenshot with Ping sketch running on the dev board.*

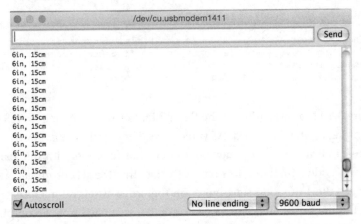

Figure 3-11   *Serial Monitor screenshot with a Ping sketch running on an Arduino Uno.*

Intel version does not work with the Arduino I, as had been mentioned in an earlier chapter. Figure 3-11 is a screenshot of the Serial Monitor with a book located exactly six inches from sensor transducer faces.

There are likely other sensors and similar devices that will not function with the dev board due to its critical timing deficiencies. This is really not much of an issue, as the Edison is very fast when used with its breakout board and programmed in Python. This method is discussed later in the book and will readily meet any reasonable timing requirements that devices like a ping sensor need. You just can't use the Arduino emulation for devices needing microsecond timing requirements.

Summary

The gist of this chapter was to provide some basic instructions on how to use the Processing language with an Intel Arduino Development Board, using the Intel Arduino IDE. Several practical examples were provided to help clarify the instructions. In addition, some key limitations of the Intel Arduino emulation were pointed out, which readers should keep in mind as they build projects using this particular environment.

4

Edison-Controlled Robotic Car

In this chapter, I will show you how to build a small robotic car that is controlled by the Intel Edison Arduino Development Board (dev board). Some of the concepts and sketches discussed in Chapter 3 will be put to practical use in this project.

BOE-BOT Car

The robotic car used in this project is built from a Board of Education Robot (BOE-BOT) kit, part number 32335, purchased from Parallax, Inc. They provide several kit versions supporting different microprocessor boards. This particular kit contains an Arduino Shield that allows for easy interface construction between the various drive components and the dev board, which is the controller for this project. Figure 4-1 shows the Arduino Shield, which mounts directly to the dev board.

The shield board simply extends the dev board pins and provides a small solderless breadboard, which allows for easy external component wiring. The shield also has a convenient power switch, which is lacking on the dev board.

I also replaced the five-pack AA battery supply that is provided in the kit with a LiPo battery/charger board with batteries, part number 28989, also available from Parallax. This battery board was initially discussed in Chapter 3 and is shown in Figure 4-2.

This particular battery board provides 7.4 V at a maximum of 2 A, which more than adequately meets the power source requirements for this project. The main

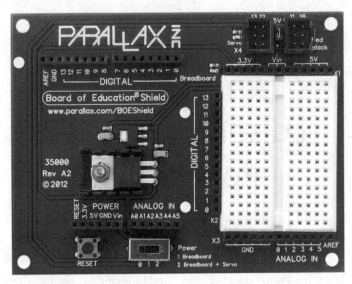

Figure 4-1 *Arduino Shield.*

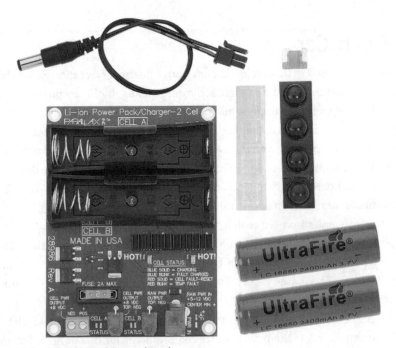

Figure 4-2 *LiPo battery/charger board.*

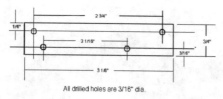

All drilled holes are 3/16" dia.

Figure 4-3 *Mounting plate adapter.*

power consumers are two continuous rotation servos that require about 250 ma each, when operating at full speed. The dev and shield boards each have their own voltage regulators that provide the proper voltages required for all board mounted components.

One minor problem in building this car by using the dev board is that the mounting-hole placements on the dev board differ from those on the LiPo battery board, which happen to perfectly match the mounting holes on the car chassis. I managed to accommodate the hole differences by crafting an acrylic mounting adapter. A dimension drawing of this adapter is shown in Figure 4-3.

You will also need some additional spacers and 4-40 screws beyond what is contained in the original kit. I have described these additional parts in Table 4-1.

You will need to follow the basic BOE_BOT assembly instructions, which are located on the Parallax website. I would recommend that you first mount the dev board to the LiPo board before you mount the LiPo board to the main chassis. You will need some free space to attach the mounting screws and nuts that hold these boards together. Other than that hint, the rest of the assembly follows the Parallax instructions, which I found to be quite clear and unambiguous. The complete assembly is shown in Figure 4-4.

You should note that at this point in the project there are no sensors attached to the car. I will add some in a later section, but at this stage, I just wanted to assemble and test a very basic car platform. Now it is appropriate to explain how

Part	Qty	Description	How Used
Spacer	4	1/2" Nylon, non-threaded	Mount LiPo board
Screw	4	4-40, 1 c long	Mount dev board
Nut	4	4-40	Mount dev board
Plate	1	1/8" Acrylic sheet, 4" × 10"	Adapter plate

Table 4-1 *Additional Hardware Required for the Robotic Car Assembly*

Figure 4-4   *Assembled robot car.*

an analog servo functions, as it will be hard to understand how the controlling sketch works without this background information.

How an Analog Servo Works

Figure 4-5 is a somewhat transparent view of the inner workings of a standard analog R/C servo motor. I would like to point out five components in this figure:

1. *Brushed electric motor*—left side
2. *Gear set*—just below the case top
3. *Servo horn*—attached to a shaft protruding above the case top
4. *Feedback potentiometer*—at the bottom end of the same shaft with the horn
5. *Control PCB*—bottom of the case to the motor's right

The electric motor is just an inexpensive, ordinary motor that probably runs at approximately 12,000 r/min unloaded. It typically operates in the 2.5- to 5-V DC

Figure 4-5 *Inner view of a standard R/C servo motor.*

range and likely uses less than 200 ma, even when fully loaded. The servo torque advantage results from the motor spinning the gear set such that the resultant speed is reduced significantly, resulting in a very large torque increase, as compared to the motor's ungeared rating. A typical motor used in this servo class might have a 0.1-oz-in torque rating, while the servo output torque could be about 42 oz-in, which is a 420 times increase in torque production. Of course, the speed would be reduced by the same proportional amount, going from 12,000 r/min to about 30 r/min. This slow speed is still sufficiently fast enough to move the servo shaft to meet normal R/C requirements.

The feedback potentiometer attached to the bottom of the output shaft is a key element in positioning the shaft in accordance with the pulses being received by the servo electronic control board. You may clearly see the feedback potentiometer in Figure 4-6, which is another image of a disassembled servo.

Figure 4-6 *Disassembled servo showing the feedback potentiometer.*

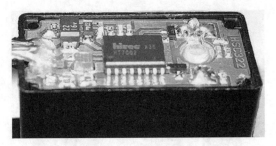

Figure 4-7   *Hitec HS-311 electronics board.*

I will discuss the potentiometer's function further during the control board analysis. The electronics board is the heart of the servo and controls how the servo functions. I will be describing an analog control version, since that is, by far, the most popular type used in low-cost servo motors. Figure 4-7 shows a Hitec control board that is in place for its model HS-311, which is a very common and inexpensive analog servo.

The main chip is labeled as HT7002, which is a Hitec private model number, as well as I could determine. I believe this chip functions the same as a commercially available chip manufactured by Mitsubishi with a model number of M51660L. I will use the M51660L as the basis of my discussion because it is used in a number of other manufacturer's servo motors and would be representative of any chip that is used in this situation. The Mitsubishi chip is entitled a "Servo Motor Controller for Radio Control," and its pin configuration is shown in Figure 4-8.

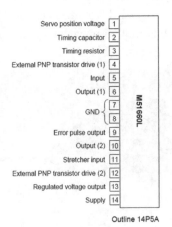

Servo position voltage — 1
Timing capacitor — 2
Timing resistor — 3
External PNP transistor drive (1) — 4
Input — 5
Output (1) — 6
GND — 7, 8
Error pulse output — 9
Output (2) — 10
Stretcher input — 11
External PNP transistor drive (2) — 12
Regulated voltage output — 13
Supply — 14

M51660L

Outline 14P5A

Figure 4-8   *Mitsubishi M51660L pin configuration.*

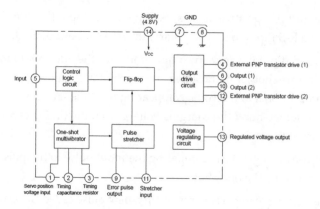

Figure 4-9 *M51660L Block Diagram.*

Don't be put off by the different physical configuration between the HT7002 in Figure 4-7 and the chip outline in Figure 4-8, as it is often the case that identical chip dies are placed into different physical packages for any number of reasons. The M51660L block diagram shown in Figure 4-9 illustrates the key functional circuits incorporated into this chip.

I will next provide an analysis that will go hand-in-hand with the Figure 4-10 demonstration circuit that was provided in the manufacturer's datasheet, as were the previous two figures.

This analysis should help you understand how an analog servo functions and why there are certain limitations inherent in its design.

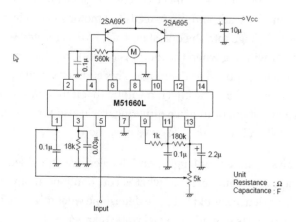

Figure 4-10 *Demonstration M51660L schematic.*

1. The start of a positive pulse appearing on the input line (pin 5) turns on the R-S flip-flop and also starts the one-shot multivibrator running.

2. The R-S flip-flop works in conjunction with the one-shot to form a linear one-shot, or monostable, multivibrator circuit whose "on" time is proportional to the voltage appearing from the tap on the feedback potentiometer and the charging voltage from the timing capacitor attached to pin 2.

3. The control logic starts comparing the input pulse to the pulse being generated by the one-shot.

4. This ongoing comparison results in a new pulse called the error pulse, which is then fed to the pulse stretcher, deadband, and trigger circuits.

5. The pulse stretcher output ultimately drives the motor control circuit that works in combination with the directional control inputs that originate from the R-S flip-flop. The trigger circuits enable the PNP transistor driver gates for a time period directly proportional to the error pulse.

6. The PNP transistor drive gate outputs are pins 4 and 12, which control two external PNP power transistors that can provide over 200 ma to power the motor. The M51660L chip can provide up to 20 ma without using these external transistors. That is too small a current flow to power the motor in the servo. The corresponding current sinks (return paths) for the external transistors are pins 6 and 10.

7. The 560-kΩ resistor (R_f), connected between pin 2 and the junction of one of the motor leads and pin 6, feeds the motor's back electromotive force (EMF) voltage into the one-shot. Back EMF is created within the motor stator winding when the motor is coasting or when no power pulses are being applied to the motor. This additional voltage input results in a servo damping effect, meaning that it moderates or lessens any servo overshoot or in-place dithering. I will also further discuss the R_f resistor when I cover the CR servo operation.

The above analysis, while a bit lengthy and detailed, was provided to give you an understanding of the complexity of what is constantly occurring within the servo case. This knowledge should help you determine what might be happening if one of your servos starts operating in an erratic manner.

There was the word "deadband" mentioned in step 3 that is worth some more explanation. Deadband used in this context refers to a slight voltage change in the control input that should not elicit an output. This is a deliberate design feature to prevent the servo from reacting to any slight input changes. Using a deadband improves servo life and makes it less jittery during normal operations. The deadband is fixed in the demonstration circuit by a 1-kΩ resistor connected between pins 9 and 11. This resistor forms another feedback loop between the pulse stretcher input and output.

The last servo parameter I will discuss is the pulse stretcher gain, which largely controls the error pulse length. This gain in the demonstration circuit is set by the values of the capacitor from pin 11 to ground and the resistor connected between pins 11 and 13. This gain would also be referred to as the proportional gain (K_p) in closed-loop control theory. It is important to have the gain set to what is sometimes jokingly called the "Goldie Locks region", not too high nor too low, but just right. Too much gain makes the servo much too sensitive and possibly could lead to unstable oscillations. Too little gain makes it too insensitive and causes a very poor time response. Sometimes, experimenters will tweak the resistor and capacitor values in an effort to squeeze out a bit more performance from a servo; however, I believe the manufacturers have already set the component values for a good balance between performance and stability.

There is one more important servo topic to discuss, namely, the continuous rotation servo, which is the type used in the robotic car.

Continuous Rotation (CR) Servos

Sometimes you will need a servo to act as a normal motor with the added advantage of being able to closely control both the speed and rotation direction. I have used continuous rotation (CR) servos for quite a long time in the robots I build for both classroom and personal use. You may easily purchase CR servos, or you can fairly easily convert a standard servo to a CR type. CR servos are almost identical in price to standard servos. I will explain the difference between the two, and you can decide if you want to convert or purchase.

The standard analog servo has a mechanical stop in place on a gear that is part of the main output shaft. This tab restricts the output shaft to a fixed range of motion, usually 180°. Figure 4-11 shows this mechanical stop on a standard servo gear train set.

Figure 4-11 *Mechanical stop in a standard servo.*

I would recommend trying to snap the tab off with a sharp diagonal cutter rather than filing it down. Be sure you disassemble the gear set before working on it because you don't want any plastic shards or filings gumming up the gear train. Figure 4-12 shows the tab neatly removed and filed flat.

The next step in the conversion process is to remove the potentiometer by desoldering it from the circuit board. The potentiometer also has built-in stops, which would restrict the output shaft if it were not removed. The potentiometer must be replaced with a resistor divider circuit that supplies the mid-point voltage to the one-shot multivibrator. Figure 4-13 shows an altered demonstration schematic with two 2.2-kΩ resistors replacing the potentiometer.

Now the control chip believes it is always at the center point; and when you supply an input pulse waveform with a width greater than 1.5 ms, the controller

Figure 4-12 *Tab removed from gear.*

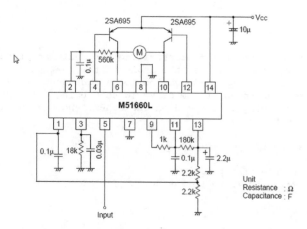

Figure 4-13 *Demonstration M51660L schematic altered for CR operation.*

will drive the motor in a CW direction. Conversely, if the input pulse width is less than 1.5 ms, it will drive the motor in a CCW direction. Additionally, as you decrease or increase the pulse width, the motor will rotate faster in the respective direction. This means 2.0 ms produces the maximum speed in the CW direction, while 1 ms produces the maximum speed in the CCW direction.

The only disadvantage is that the motor will tend to creep if your resistor divider doesn't produce exactly the mid-point voltage. Exactly how much is hard to predict because the torque demands play a part in actually moving whatever object is being powered by the CR servos. A large robot would likely not even move due to the minute creep signal created. I would definitely use matched or precision resistors in order to divide the voltage as precisely as possible.

Another way to address this issue is to alter the value of the deadband resistor (the 1 kΩ) to help eliminate the undesired motion. It would be a trial and error process to determine the correct value.

There is one final precaution that you should know. It is entirely possible that the plus or minus 0.5 ms deviation from the center 1.5-ms pulse width will not produce the full rotation speed change that is possible. This is entirely due to having too large a value for the feedback resister R_f. The value set for this resistor in the demonstration circuit is 560 kΩ. This may have to be lowered to 120 kΩ to achieve the full speed capability for pulse widths that range from 1.0 to 2.0 ms.

Creating a sketch that controls CR servos requires an understanding of the Servo class `write` method. I have included the following excerpt from the Servo API that describes how to use the `write` method with a CR servo.

```
Servo
write()
Description
Writes a value to the servo, controlling the shaft accordingly. On a
standard servo, this will set the angle of the shaft (in degrees),
moving the shaft to that orientation. On a continuous rotation servo,
this will set the speed of the servo (with 0 being full speed in one
direction, 180 being full speed in the other, and a value near 90 be-
ing no movement).
Syntax
servo.write(angle)
Parameters
servo: a variable of type Servo
angle: the value to write to the servo, from 0 to 180
Example
#include <Servo.h>

Servo myservo;

void setup()
{
  myservo.attach(9);
  myservo.write(90);   // set servo to mid-point
}

void loop() {}
```

Servo1 Sketch

My initial sketch is named Servo1 and will drive the robot car in a forward direction at the maximum speed, which isn't too fast as you will discover when you operate the car. I covered about 32 feet in a minute, which is approximately 46 r/min, given the 2.5-inch wheel diameter.

```
// Servo1
// by D. J. Norris 1/2015
// This example code is in the public domain.

#include <Servo.h>

int maxSpeedFwd = 0;
int maxSpeedBack = 180;

Servo myservo1,myservo2; // create two servo objects
                         // maximum of eight servo objects can be
                            created

void setup()
{
  myservo1.attach(3);  // attaches a servo on pin 3 to the myservo1
                          object
```

```
myservo2.attach(9);   // attaches a servo on pin 9 to the myservo2
                         object
}

void loop()
{
    myservo1.write(maxSpeedBack);
    myservo2.write(maxSpeedFwd);
}
```

I also want to mention why I used pins 3 and 9 to drive the servos in lieu of pins 12 and 13, which would normally be used for the sketch, since it was written for the Parallax kit using an Arduino Uno controller. The dev board does not appear to support analog servo control on any pins other than the pins I selected. I am unsure why this is the case, but I suspect it must be related to the underlying Edison GPIO pins that are brought out to the dev board shield pins. Just be aware that this is just another issue with using the dev board versus a real Arduino. And while addressing compatibility issues, I would also point out that the dev board does not appear to directly support the `writeMicroseconds(<number of microseconds>)` method, which is part of the Servo class. Again, this lack of functionality is likely related to the critical timing that I discussed earlier. Not being able to use the `writeMicroseconds` method just means that most of the example servo sketches will not work with the dev board. They can be loaded into the board, but they will not function correctly.

The Servo `write` methods do appear to work correctly as I have demonstrated with my sample Servo1 sketch. I have also included Figure 4-14, which is a screenshot of the servo pulse waveforms going to both CR servos.

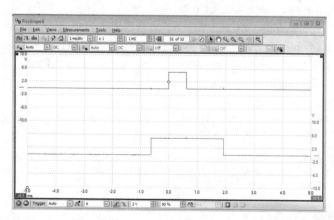

Figure 4-14   *Servo waveforms.*

One pulse is 0.5 ms and the other is 2.5 ms, which, according to the API, will drive CR servos to maximum speeds in opposite rotational directions. This allows the robot car to move forward at its maximum speed, since the servos are mounted as mirror images.

Having now presented a working robot car, I will next show how to incorporate an obstacle sensor to provide a bit of autonomous operation to the car.

Autonomous Operation

Autonomous, or *stand-alone,* operation allows a robot to operate without direct user intervention. I will accomplish this by adding an ultrasonic sensor to the robot car that will be able to detect nearby obstacles and alert the robot controller, which is the dev board for this project. The ultrasonic sensor will be the same ping sensor discussed in Chapter 3. I will need to include a dedicated Arduino Uno to operate this sensor because the dev board does not operate it, as I discussed in Chapter 3. The Uno will be programmed to toggle a GPIO line if an obstacle is sensed within 10 inches or closer to the sensor transducers. The dev board is programmed to poll, or repeatedly test, this line to see if it has transitioned to a HIGH state. If it has, the sketch controlling the car will enter an obstacle avoidance routine, which, hopefully, will cause the robot to go on its way without a problem. This coding is actually a simple artificial intelligence (AI) approach that adds some autonomy to the robot's behavior. Thus, the AI relieves a human operator from directly controlling the car, as would be the case for a typical radio-controlled (RC) vehicle.

Figure 4-15 is a complete wiring diagram for the robot car, which you can use as a guide when you wire your own car.

Next, I need to show you the sketch that is loaded into the Arduino Uno that will trigger an output on its pin 2 if an obstacle is located 10 inches or less in front of the sensor. The sketch is named Ping1, as it is a modified Ping sketch contained in the Arduino Examples library.

```
/* Ping))) Sensor
   This sketch reads a PING))) ultrasonic rangefinder and returns the
   distance to the closest object in range. To do this, it sends a
   pulse
   to the sensor to initiate a reading, then listens for a pulse
   to return.  The length of the returning pulse is proportional to
   the distance of the object from the sensor.

   The circuit:
     * +V connection of the PING))) attached to +5 V
```

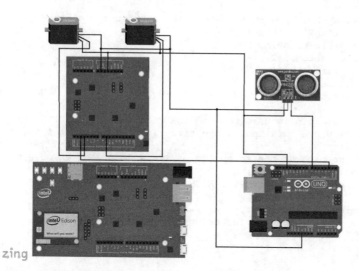

Figure 4-15 *Robot car wiring diagram.*

```
     * GND connection of the PING))) attached to ground
     * SIG connection of the PING))) attached to digital pin 7

  http://www.arduino.cc/en/Tutorial/Ping

  created 3 Nov 2008
  by David A. Mellis
  modified 30 Aug 2011
  by Tom Igoe
  further modified by D. J. Norris 1/2015
  This example code is in the public domain.
  */

// these constants won't change. It's the pin number
// of the sensor's output and signaling line to the dev board
const int pingPin = 7;
const int signalPin = 2;

void setup() { // nothing required for the setup method
}

void loop()
{
  // establish variables for duration of the ping,
  // and the distance result in inches
  long duration, inches;

  // The PING))) is triggered by a HIGH pulse of 2 or more
     microseconds.
```

```
// Give a short LOW pulse beforehand to ensure a clean HIGH pulse:
pinMode(pingPin, OUTPUT);
digitalWrite(pingPin, LOW);
delayMicroseconds(2);
digitalWrite(pingPin, HIGH);
delayMicroseconds(5);
digitalWrite(pingPin, LOW);

// The same pin is used to read the signal from the PING))): a HIGH
// pulse whose duration is the time (in microseconds) from the
   sending
// of the ping to the reception of its echo off of an object.
pinMode(pingPin, INPUT);
duration = pulseIn(pingPin, HIGH);

// convert the time into a distance
inches = microsecondsToInches(duration);

if(inches < 10)
{
  pinMode(signalPin, OUTPUT);
  digitalWrite(signalPin, HIGH);
  delay(10);
}
digitalWrite(signalPin, LOW);
delay(100);
}

long microsecondsToInches(long microseconds)
{
  // According to Parallax's datasheet for the PING))), there are
  // 73.746 microseconds per inch (i.e. sound travels at 1130 feet per
  // second). This gives the distance travelled by the ping, outbound
  // and return, so we divide by 2 to get the distance of the
     obstacle.
  // See: http://www.parallax.com/dl/docs/prod/acc/28015-PING-v1.3.pdf
  return microseconds / 74 / 2;
}
```

The previous sketch, Servo1, also needs to be modified so that it constantly checks for the signal transition from the Arduino Uno indicating an obstacle is detected. I elected to use pin 2 on the dev board as an input for the obstacle detection module. The modified Servo1 code is shown below. I also renamed it to Servo2 to differentiate from the original sketch.

```
// Servo2
// by D. J. Norris 1/2015
// This example code is in the public domain.

#include <Servo.h>
```

```
int maxSpeedFwd = 0;
int maxSpeedBack = 180;

// need a slower speed for turning
int medSpeedFwd = 45;

// need a zero speed value
int zeroSpeed = 90;

// the obstacle signal pin from the Uno
int signal = 2;

// variable to hold detection state
int obstacle;

Servo myservo1,myservo2; // create two servo objects
                         // maximum of eight servo objects can be
                            created

void setup()
{
  myservo1.attach(3);  // attaches a servo on pin 3 to the myservo1
                          object
  myservo2.attach(9);  // attaches a servo on pin 9 to the myservo2
                          object
  pinMode(signal, INPUT); //set up the input from the Arduino Uno
}

void loop()
{
    myservo1.write(maxSpeedBack);
    myservo2.write(maxSpeedFwd);

// check for an obstacle
obstacle = digitalRead(signal);
  if(obstacle) {
      avoid();  // call the obstacle avoidance method
  }
}

void avoid()
{
    myservo1.write(medSpeedFwd); // these two commands turn bot to
                                    the left
    myservo2.write(zeroSpeed);
    delay(1000);  //turns about 45 degrees in one second
}
```

An oblique front view of the fully assembled car is shown in Figure 4-16. The Arduino Uno may be clearly seen mounted to the top of the Arduino shield with

Figure 4-16 *Front view of the fully assembled robot car.*

three 1-in aluminum spacers. The shield and Uno each have three matching holes, which makes the mounting very easy.

The ping sensor is plugged into a small, solderless breadboard, which is also mounted on an L-shaped piece of lightweight gauge aluminum stock. This bracket is attached to the car chassis by a pair of 3/8" 4-40 machine screws and nuts.

Figure 4-17 is a view of the rear of the car. You may notice a lead fishing sinker attached with nylon tie wraps to the back edge of the car chassis. I had to add some extra weight to the car's back due to it being a bit unbalanced and tending to drop on its front side.

Operating the Robot Car

You start the car by simply sliding the shield slide switch from the 0 to the 1 position. You do not need to use the slide 2 position because the CR servos are connected to 5 V and not to Vin. After a delay of approximately 15 seconds, the car

Figure 4-17   *Rear view of the car.*

will move forward until it encounters an obstacle in its direct path. The car will then turn left about 45° and then attempt to go forward. If an obstacle is still detected, the car will turn further left another 45°. This should now normally provide a clear path forward for the car.

Summary

I showed you how to build, program, and operate an autonomous robotic car using the dev board as the main controller. Certain constraints and limitations of the dev board were also clearly identified in this chapter. I also provided a fairly extensive discussion on how an analog servo functions to help you understand the CR servos used with this car.

Now it is time to show you how to program the Edison using a totally different approach.

5

Connecting to Edison Linux with the Command-Line Prompt

In this chapter, I will show you how to connect to an Edison module mounted on the Intel Edison Breakout Board, which I will now simply refer to as the breakout board. I will be using a Macbook Pro running a terminal program as my client to the breakout board. I will also demonstrate some basic Linux commands sent to the Edison, which is running the default Poky Linux distribution.

Intel Edison Breakout Board

I first showed you the breakout board in Chapter 1, but it did not have an Edison mounted on it. Figure 5-1 is a picture of the board with an Edison module mounted on it.

Looks can be deceiving, and the breakout board has quite a bit of functionality built into it despite its small size. It has a Hirose connector on which you mount the Edison module. It also has two micro USB connectors: one marked J3 for client USB communications and the other one marked J16 for the USB OTG and power connections. I previously discussed an OTG connection in Chapter 2, so I will not repeat that discussion here. The client USB connector is the one nearer the center of the right edge of the board, just above a jumper marked as J21. You will need two micro USB cables plugged into the breakout board: one

Figure 5-1   *Intel Edison Breakout Board.*

for power and the other for USB serial communications. You are all set for a connection session, once you have the two cables connected between the breakout board and the laptop.

Setting Up Your First USB Communications Session

As mentioned earlier, I used a Macbook Pro as a client to connect with the Edison. The default Terminal program on the Macbook served quite well as a client. If you are using a Windows or Linux computer, I would suggest using a free, open-source terminal program, such as TeraTerm. There are other similar applications available that I am sure would also function well as a client program.

FTDI Drivers

For your first connection, you will need to first install the FTDI drivers, which permit the client computer (the one running the terminal program) to connect to the host (the Edison). The host uses an FTDI chip to implement the USB communications, and these drivers establish virtual ports on the client. These virtual ports will be shown as Comm xx in the Windows Device Manager Ports section. The xx in the name will depend upon the number and type of comm ports already installed on the client computer. In a Mac computer, they will be identified as either `tty.usbserial` or `cu.usbserial` followed by a long alphanumeric identifier. I discuss the Mac connection in more detail in a following section.

These FTDI drivers may be downloaded from the Intel Edison downloads web page. As of the time of this writing, the driver is named CDM v2.10.00 and is approximately 2 MB in size.

Windows Drivers

Windows installations also require some additional drivers to implement RNDIS, CDC, and DFU. These drivers are also available on the Intel Edison downloads page. Just look for the selection entitled Windows Driver Setup. Click on this selection, and a file named IntelEdisonDriverSetup1.0.0.exe will be downloaded to your computer. Just execute this file, and it will automatically complete the driver installations.

I have provided a brief description of these drivers below for those readers who might be interested.

- *Remote Network Driver Interface Specification* (RNDIS)—This driver creates a virtual Ethernet link over a physical USB connection. This link type is required to establish the initial USB communications with the Edison module.

- *Composite Device Class* (CDC)—This is a supporting class for RNDIS and is required for USB device recognition and communications.

- *Device Firmware Upgrade* (DFU)—This is an application that is used to update and upgrade the Edison's firmware for Windows systems.

Connecting to the Client Computer

You will initially need to identify the logical device name that the Edison establishes when first connected. Use the following command to do this:

```
ls /dev/tty.*
```

Figure 5-2 is a terminal screenshot for the results of this command. The name you will look for in a Mac OS X environment contains either `cu.usbserial` or `tty.usbserial`. The entry `/tty.usbserial-A502LTX2` was exactly what I needed.

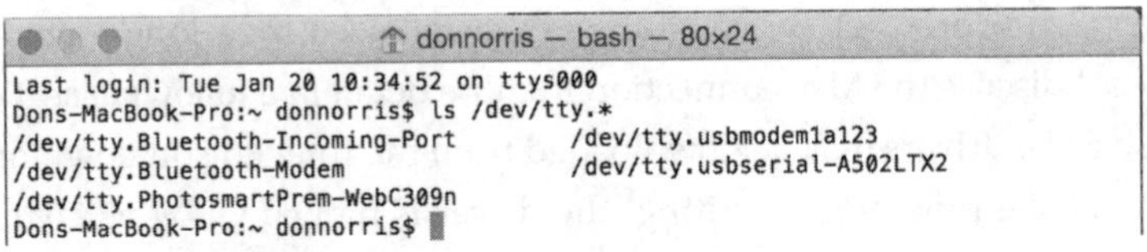

```
●  ●  ●                ⌂ donnorris — bash — 80×24
Last login: Tue Jan 20 10:34:52 on ttys000
Dons—MacBook—Pro:~ donnorris$ ls /dev/tty.*
/dev/tty.Bluetooth-Incoming-Port      /dev/tty.usbmodem1a123
/dev/tty.Bluetooth-Modem              /dev/tty.usbserial-A502LTX2
/dev/tty.PhotosmartPrem-WebC309n
Dons—MacBook—Pro:~ donnorris$ ▊
```

Figure 5-2 *Terminal screenshot displaying serial USB host names.*

I next entered the following to establish the actual communications link:

```
screen /dev/tty.usbserial-A502LTX2 115200 -L
```

The screen portion creates a session in the terminal application with the logi-cally named device `tty.usbserial-A502LTX2` using a 115,200 baud rate. The `-L` at the end creates an automatic log file that records all the session activity. This log file is very useful for troubleshooting problems that might arise with the com-munications link. I had to press the enter key twice to get the link working. This is due to an annoying problem with the Edison in which it will "go to sleep" after five seconds of inactivity on the USB serial line. Not a major issue; just an annoyance.

Figure 5-3 is a terminal screenshot of my second login with the Edison. I inad-vertently terminated my first one by disrupting the power to the board

You must enter `root` when prompted for a username and then simply press the return key when prompted for the password. You are now connected via the USB to the breakout board. The command-line prompt shown is:

```
root@edison:~#
```

This prompt shows that the current connection is to the home directory (indicated by the tilde [~] shown immediately after the colon) of a user named `root` who is active on a Linux OS device named Edison. You should note that using `root` as a user will pose a certain security risk because `root` has access

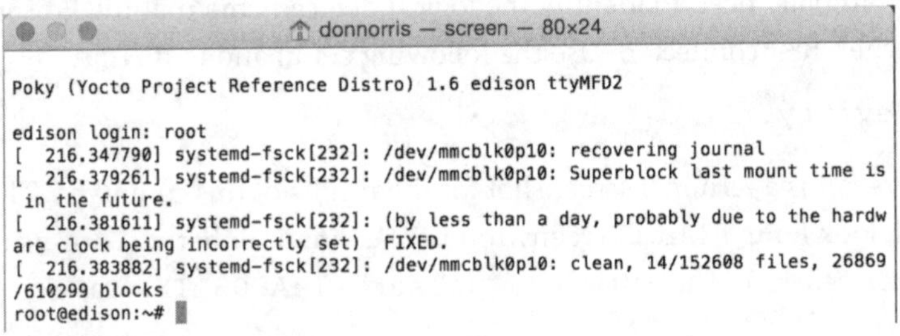

```
●  ●  ●                ⌂ donnorris — screen — 80×24
Poky (Yocto Project Reference Distro) 1.6 edison ttyMFD2

edison login: root
[  216.347790] systemd-fsck[232]: /dev/mmcblk0p10: recovering journal
[  216.379261] systemd-fsck[232]: /dev/mmcblk0p10: Superblock last mount time is
in the future.
[  216.381611] systemd-fsck[232]: (by less than a day, probably due to the hardw
are clock being incorrectly set)  FIXED.
[  216.383882] systemd-fsck[232]: /dev/mmcblk0p10: clean, 14/152608 files, 26869
/610299 blocks
root@edison:~# ▊
```

Figure 5-3 *Terminal login screenshot.*

and permissions to everything in the Linux OS. Anyone who maliciously takes over as root has access to all files, passwords, etc. and could cause havoc with your system.

The next section shows you how to update/upgrade the Edison's firmware.

Updating/Upgrading the Edison Firmware

I have titled this section Updating/Upgrading the Edison firmware because the existing Edison documentation uses both terms to indicate that users should download and install the latest version of the module's firmware. I totally agree, as improvements to the firmware are created almost weekly. In fact, you will not be able to replicate some of the software examples that follow without installing a "fresh" firmware package.

The process of updating, as I will call it from now on, is relatively simple. There should be a new folder named Edison, which is displayed on the client computer. Initially, this folder will be empty. This folder is a result of the USB driver installation. It iconically identifies the Edison as a mass storage device to the client. It is also the memory location that the existing Edison software will access to update its own firmware.

You will need to download the latest firmware update from the Intel Edison downloads web page. Look for the selection named "Edison Yocto complete image." It is about a 105-MB zipped file that then needs to be extracted to a set of 26 files and one directory of approximately 608 MB in size. This extracted package of files and one directory should then be copied into the empty Edison folder. Now you are ready to update the Edison. To do so, type the following at the command-line prompt:

```
reboot ota
```

The Edison will first shut down and then begin a restart during which it will load the new Linux image from the Edison folder. Be patient and do not interrupt the process, which should take several minutes to complete. After the load completes, the same command-line prompt should appear as discussed above. Just enter root for the user and press the enter key if prompted for a password.

Next you should run the built-in configuration application named configure_ edison, which will allow you to create a new host name, create a password, and start and configure the WiFi.

configure_edison Application

To start this configuration application, enter the following at the prompt and follow all the subsequent on-screen instructions:

```
configure_edison
```

Figure 5-4 is a screenshot of the WiFi configuration portion of the setup. I found it very easy to set up the Edison's WiFi. The Edison appears to have a sensitive and well-integrated WiFi module that comes with all the driver software preinstalled. You should note the IP address assigned to the Edison, as you will need it to confirm that the Edison's web server is functioning properly. The local IP address assigned for my installation was 192.168.0.9. Obviously, your address will differ, but it makes no difference other than what you enter in a browser's URL line.

The Edison's name, password, or WiFi configuration may be set individually at the command-line prompt by entering configure_edison followed by --name, --password, or --wifi, respectively. In addition, you can always get the help screen by entering:

```
configure_edison --help
```

Figure 5-5 is a screenshot of the configure_edison application with the help option inserted in the command line.

```
●  ●  ●              ⬆ donnorris — screen — 80×24
Configure Edison: WiFi Connection

Scanning: 1 seconds left

0 :     Rescan for networks
1 :     Manually input a hidden SSID
2 :     xfinitywifi
3 :     Carpentier
4 :     \x00\x00\x00\x00\x00\x00\x00\x00\x00\x00\x00\x00
5 :     HOME-5395
6 :     MOTOROLA-5A991


Enter 0 to rescan for networks.
Enter 1 to input a hidden network SSID.
Enter a number between 2 to 6 to choose one of the listed network SSIDs: 6
Is MOTOROLA-5A991 correct? [Y or N]: y
What is the network password?: ********************
Initiating connection to MOTOROLA-5A991...
Done. Network access should be available shortly, please check 'wpa_cli status'.
Connected. Please go to 192.168.0.9 in your browser to check if this is correct.
root@edison1:~# configure_edison --server
Connected. Please go to 192.168.0.9 in your browser to check if this is correct.
root@edison1:~# █
```

Figure 5-4 *WiFi configuration screenshot.*

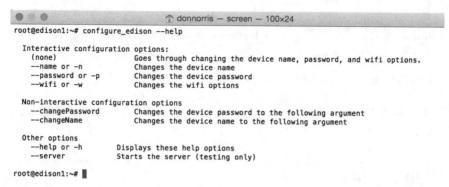

```
root@edison1:~# configure_edison --help

    Interactive configuration options:
       (none)              Goes through changing the device name, password, and wifi options.
       --name or -n        Changes the device name
       --password or -p    Changes the device password
       --wifi or -w        Changes the wifi options

    Non-interactive configuration options
       --changePassword    Changes the device password to the following argument
       --changeName        Changes the device name to the following argument

    Other options
       --help or -h        Displays these help options
       --server            Starts the server (testing only)

root@edison1:~#
```

Figure 5-5   *Configure_edison help option screenshot.*

Date and Time

The Edison updates its internal clock by attempting to contact a public network time protocol (NTP) server. You must have Internet access for this to work because there is no internal real-time clock in the Edison module or on any of the development boards. Figure 5-6 shows the result of entering the date command.

In this figure, the date is correct but the time is off by three hours, as the default NTP server was apparently using a non-local time zone. You can change the local time zone by following these steps:

1. Delete the existing local timezone link by entering:
   rm /etc/localtime

2. Create a link to the appropriate timezone, which is found in the file /usr/share/zoneinfo/<your region>/<major city in timezone>. In my case, the region is America and the city is New_York.
   ln -s /usr/share/zoneinfo/America/New_York /etc/localtime

3. List the new link just to be certain it was created correctly.
   ls -l /etc/localtime

4. Reboot the Edison to make the link active.
   reboot

```
root@edison1:~# date
Tue Jan 20 17:53:18 UTC 2015
root@edison1:~#
```

Figure 5-6   *Date command result.*

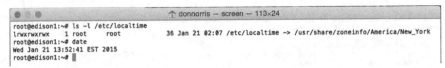

Figure 5-7   *Timezone* link and corrected date/time.

Figure 5-7 shows the timezone link and the correct time and date after a reboot.

Web Server

The Edison firmware has a web server preinstalled, but I found that it needs to be started. You will need to enter the following at the command-line prompt in order to start the web service:

```
configure_edison --server
```

Next, enter the IP address that was previously assigned to the Edison in your client computer's browser. You should see the web page shown in Figure 5-8. This page shows only the host name, which should be edison if you haven't changed it from the default of the configure_edison application. The current, local IP address should also be displayed.

NOTE   If you did not update the firmware, then the web page displayed will be completely different. The original web page provided a GUI view of the configure_edison application, allowing you to change the device name and password as well as reconfigure the WiFi. This web page is shown in Figure 5-9

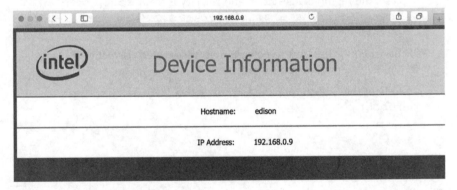

Figure 5-8   *Edison web server page.*

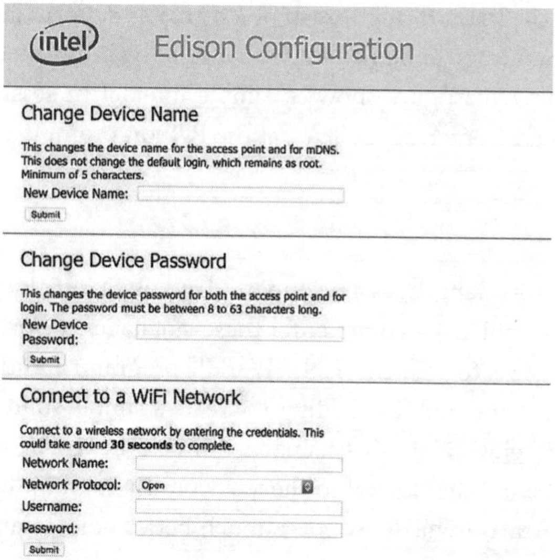

Figure 5-9 *Original Edison web server page.*

for your information only, but you shouldn't ordinarily see it because you need to be using the most current firmware to get the most utility from the Edison.

Now it is time to examine some of the preinstalled language applications including Python, gcc, and Node.js.

Python

Python version 2.7.3 comes preinstalled with the Poky Linux distribution, as shown in Figure 5-10. You start it in the interactive mode by simply entering `python` at the command-line prompt.

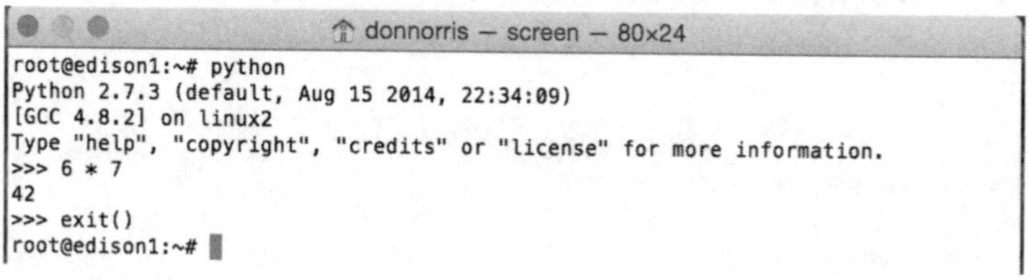

```
root@edison1:~# python
Python 2.7.3 (default, Aug 15 2014, 22:34:09)
[GCC 4.8.2] on linux2
Type "help", "copyright", "credits" or "license" for more information.
>>> 6 * 7
42
>>> exit()
root@edison1:~#
```

Figure 5-10 *Python screenshot.*

I will discuss the Python language in much greater detail in later chapters, but at this point, I just want to demonstrate that it is readily available in this initial serial connection. This figure shows a simple interactive session with Python that is terminated by entering exit() at the Python prompt.

C/C++

Both the C and C++ languages are supported in Poky Linux by the gcc, which is an open-source C and C++ compiler for these languages. A straightforward way to demonstrate gcc is to create a "Hello World!" program, and then compile and run it. I had to use the built-in vi editor to create a simple source-code example, which is shown below. Note that using vi can be an exercise in frustration, as there seems to be nothing logical in the way you use it and it has no GUI functions. But take heart, I will show you a much better editor named nano in the next chapter.

```
#include <stdio.h>

main()
{
    printf("Hello World!"\n);
}
```

I saved this source code to a file named test.c and proceeded to compile it using the following command:

```
gcc -c test test.c
```

The first test in the above command is the name to be applied to the compiled code. Obviously, the second name, test.c, is the source. Once compiled, the test file can be directly executed by entering:

```
./test
```

Figure 5-11 shows the result of the compilation and program execution.

```
root@edison1:~# gcc -o test test.c
root@edison1:~# ./test
Hello World!
root@edison1:~#
```

Figure 5-11 *test.c code compiled and executed.*

Node.js

Node.js is a very interesting language that comes preinstalled in the Edison's Poky Linux distribution. It is a recently developed computer language, based mostly on Javascript, that you can use to develop both server and client-side web applications. And like Python, it also has an interactive mode, in which it is easy to demonstrate some simple aspects of the language. Of course, the easiest is just to query and display the current Node.js version, which is done by entering the following:

```
node -v
```

The response to this query is shown in Figure 5-12.

I also wanted to demonstrate a very concise "Hello World!" program, which normally would have required another attempt at using vi. I avoided that potentially frustrating experience by using a SFTP program named WinSCP to remotely transfer a file I created on a Windows machine to the Edison. SFTP is short for *secure file transfer protocol*, and it is an extremely easy way to load files into the Edison using a fast and encrypted protocol. Figure 5-13 is a screenshot of the WinSCP application as it was connected to the Edison via the LAN.

The program I transferred into the Edison was named Hello.js and contained only one line as shown below:

```
Console.log('Hello World from Node.js');
```

I ran this tiny Node.js program by entering the following:

```
node Hello.js
```

Figure 5-14 is the resulting display, which demonstrates that Node.js successfully ran the code.

The last Node.js example is more complex than the previous one and involves blinking an LED on the Arduino Development Board. This is the same dev board I used in Chapter 4; however, the Intel Arduino IDE will not be used to program it for this example. The dev board was updated before programming it because it

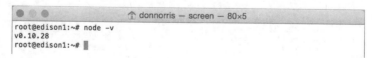

Figure 5-12 *Node.js version query response.*

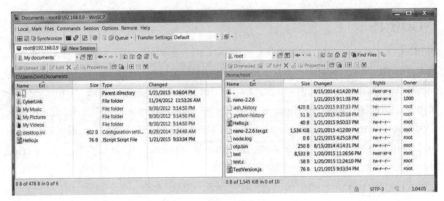

Figure 5-13 *WinSCP application connected to an Edison.*

needs a special library named mraa, which was not included in earlier firmware versions. mraa allows the Edison to have direct control over its GPIO, using a variety of languages, including Python and Node.js. The following Node.js program is named Version.js and was used to confirm that the mraa library was installed.

```
var v = require ('mraa')
console.log('mraa version is ' + v.getVersion());
```

Figure 5-15 shows the result of running this program, which is done by entering:

```
node Version.js
```

I next wrote the Node.js vesion of the Blink program on a remote Windows computer and transferred it to the dev board using the WinSCP application, which I described earlier. The Node.js version of the blink program is listed below:

```
/*
blink.js
D. J. Norris 1/2015
Code is in the public domain
This program blinks an LED attached to pin 13 on the Intel Arduino Dev
Board
```

```
                    ⬆ donnorris — screen — 80×19
root@edison1:~# node Hello.js
Hello World from Node.js
root@edison1:~# ▮
```

Figure 5-14 *Hello.js program display.*

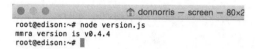

```
root@edison:~# node version.js
mmra version is v0.4.4
root@edison:~#
```

Figure 5-15   *Version.js mraa query response.*

```
Written in Node.js
*/
// load mraa and logically connect it to a var named pin
var pin = require('mraa');
// Instantiate an object named led and attach it to GPIO #13
var led = new pin.Gpio(13);
// make led an output
setup = function() {
  pinMode(led, OUTPUT);
}
// need a var to hold led state
var state = 0;
// create the blink function now
var blink = function() {
  state = (state == 1?0:1) // Tertiary function sets state to 0 if 1
and v.v.
  led.write(state);        // output the new state to pin 13
  setTimeout(blink, 500);  // recursive call every 500 ms
}
// kick it off
blink();                       // should start blinking the LED at this
                               point. Stop program with ^C
```

I entered the following to start blinking the LED on the dev board:

```
node blink.js
```

The LED did start blinking immediately and continued until I pressed the control-C key combination, which is the Linux keyboard interrupt sequence.

This example concludes my brief introduction to several of the languages used to program the Edison. I will discuss how to program the Edison using Python in the next chapter, including a much greater discussion of the mraa library.

Summary

This chapter showed you how to communicate with the Edison using USB and wireless links. I showed you how to update the module so that it executed the latest firmware. I also demonstrated several programming examples, using the Python, C/C++, and Node.js languages.

6

Debian Linux and Python Basics

In this chapter, I will expand on basic programming constructs that I first introduced in Chapter 4. This discussion will use the Python language as a learning platform and will include the concepts of data types, variables, syntax, objects, and methods. This discussion should prepare you to write your own Python programs that will enable the Edison to properly function as a project controller. However, initially I need to show you how to install the Debian Linux distribution in place of the default Poky distribution. I will state my reason for this switch in the next section.

How to Install the Debian Linux Distribution

I have chosen to replace the default Poky Linux distribution with a Wheezy Debian Linux distribution. I had three reasons to take this action:

1. It's been my experience that Debian is a much "friendlier" distribution for relatively new Linux users as compared to Poky, which is based on the developer-inspired Yocto framework. I am not denigrating Poky, but simply stating it is more focused on the engineering community than the hobbyist/enthusiast market.

2. Debian has a comprehensive package library, which allows users to easily add new applications without having to recompile and rebuild a full distribution, as might be the case with the Yocto framework, when a desired package is not available.

3. Wheezy Debian is the Linux distribution normally recommended to be used by Raspberry Pi users. I figured that many readers of this book would already be Pi users, and it would naturally make sense to take advantage of their existing familiarity with this Debian distribution.

I will be using the Intel Edison Arduino Development Board (dev board) to load the Edison module with a Wheezy Debian distribution and also to demonstrate all the Python examples. The next section is a complete step-by-step procedure to accomplish this load.

WARNING   You must follow all the steps in the order given and be particularly careful not to interfere with the actual data transfer, or you will "brick," or completely disable your Edison module.

Step-by-Step Edison Debian Load Procedure

This procedure is based on a comprehensive tutorial entitled "Loading Debian (Ubilinux) on the Edison" written by Shawn Hymel, and is available on the Learn.Sparkfun.com website. I used a Windows computer running Win7 to complete the load procedure, but there are also instructions for both Mac and Linux computers in the tutorial.

CAUTION   Make sure that you do NOT connect the dev board to the computer until instructed to do so.

1. Download and install 7-zip from the http://www.7-zip.org web page. This application will be used to extract the image and supporting utilities.

2. Download dfu-util for Windows from https://cdn.sparkfun.com/assets/ learn_tutorials/3/3/4/dfu-util-0.8-binaries.tar.xz. This utility application came from the community.spark.io where you should go if you want to learn more about it.

3. Download the Wheezy Debian image from http://www.emutexlabs.com/ ubilinux. Click on the ubilinux-for-Edison button to start the download.

4. Extract the dfu-util application using the 7-zip utility. You need to do a "double" extraction because the original file was compressed twice.

I would suggest using the default Downloads directory as a good location for all the files used in this procedure.

5. Extract the ubilinux image. It will be identified as ubilinux-edison-XXXXXX.tar.gz where the Xs are numbers that will change as revisions to the image are created. The image has also been doubly compressed so you will need to do a "double" extraction as you previously did for the dfu-util application. Ultimately, the image will be located in a directory named toFlash, which will be in the Downloads directory.

6. Copy both dfu-util.exe and libusb-1.0.dll from <your Downloads directory>\dfu-util-0.8-binaries.tar\dfu-util-0.8-binaries\ dfu-util-0.8-binaries\win32-mingw32 directory and paste them into the toFlash directory, which is also in the Downloads directory.

7. Go into the toFlash directory and double click on a batch file named flashall.bat.

8. A command window will pop up asking you to plug in the Edison board. I connected two micro-USB cables from the dev board to the Windows computer at that time.

9. The image will automatically start loading. A series of progress indicators will then appear in the command window showing that the load is proceeding. This process may take over 10 minutes to complete so be patient and DO NOT disconnect the dev board or you will likely brick the Edison module.

You will need to establish a USB serial link once the image has completed. I would suggest you use the free *putty* application for this purpose when using a Windows computer, as was the case with this procedure. *putty* may be downloaded from http://www.putty.org. You will also need to know the specific serial comm port that was created when you downloaded the virtual FTDI drivers, as was discussed in the previous chapter. As a reminder, all you have to do to find the correct communications port is to use the Windows Control Panel Device Manager and click on Ports (COM & LPT). The USB serial port for my situation was COM6.

Next, start *putty* and select Serial as the communication link type. Enter the communication port and 115200 for the baud rate. Figure 6-1 shows the *putty* configuration for these values.

Figure 6-1 putty *configuration screen.*

Click open and you should see a terminal screen from the dev board requesting a user name. Enter edison for the user name and edison again when
asked for the password. Figure 6-2 shows the resulting login to Wheezy Debian
running on an Edison.

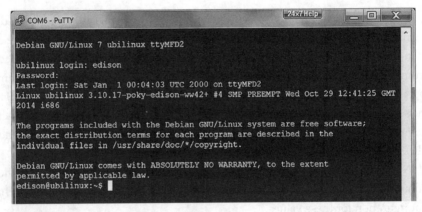

Figure 6-2 *Initial Wheezy Debian login screen.*

```
edison@ubilinux:~$ su
Password:
root@ubilinux:/home/edison# date --set="1/26/2015 21:30:00"
Mon Jan 26 21:30:00 UTC 2015
root@ubilinux:/home/edison# date
Mon Jan 26 21:30:11 UTC 2015
root@ubilinux:/home/edison#
```

Figure 6-3 *Setting date and time as the* root *user.*

Congratulations! At this point you have an active session with the Edison running a Debian distribution. Now it is time to point out a few facts regarding how to interact with this Linux distribution. Initially, if you entered a date command, it would likely show some obscure date back to 1980. You really need to be able to set the appropriate date and time in order to properly time stamp all your computer activity. You could attempt to use the Linux date command with the set option, as in the example below:

```
date --set="1/26/2015 21:30:00"
```

However, you would find that the system would not allow this operation because you do not have high enough permissions to set the time. You must login as root to accomplish this operation. Becoming the root user is easy; just type in su and then the password, which is still edison. Now as root you can set the date and time, as shown in Figure 6-3.

I will also introduce and discuss some additional Linux commands when they are needed as I proceed through the rest of the book. Next, I will show you how to set up the Edison's WiFi with this Debian distribution, since it is critical to have an Internet connection in order to gain the most utility from this Linux OS.

WiFi Setup

WiFi configuration with the Debian Linux distribution is somewhat harder to do, as compared to the Poky distribution, because Debian does not have the configure_edison utility. Instead, you must edit a networking configuration file to enable an automatic WiFi connection. There is some good news with regard to editing because Debian comes with a great text editor named nano. nano is far easier to use than vi or any of the other classic text editors that were part of earlier Linux distributions. You do have to run nano at the root level, so just enter su and the default password to become the root user. The file you have to edit is in an /etc directory, which means you should enter the following:

nano/etc/network/interfaces

There are three changes that must be made in the interfaces file in order to enable the WiFi to be automatically started at each boot up. These changes are:

1. Remove the # symbol, if one is present, before the line #auto wlan0.

2. Add your wireless network ssid after the phrase wpa-ssid.

3. Add your wireless password after the phrase wpa-psk.

You save the newly edited file by pressing the control-o key (^o) combination. You then exit the nano editor by pressing the control-x key (^x) combination. Figure 6-4 shows the unedited file.

After editing the file, you must restart the networking configuration process by entering:

/etc/init.d/networking restart

If all goes well, you should see a series of Linux script commands scroll through the console screen, eventually showing an IP address assigned as part of the

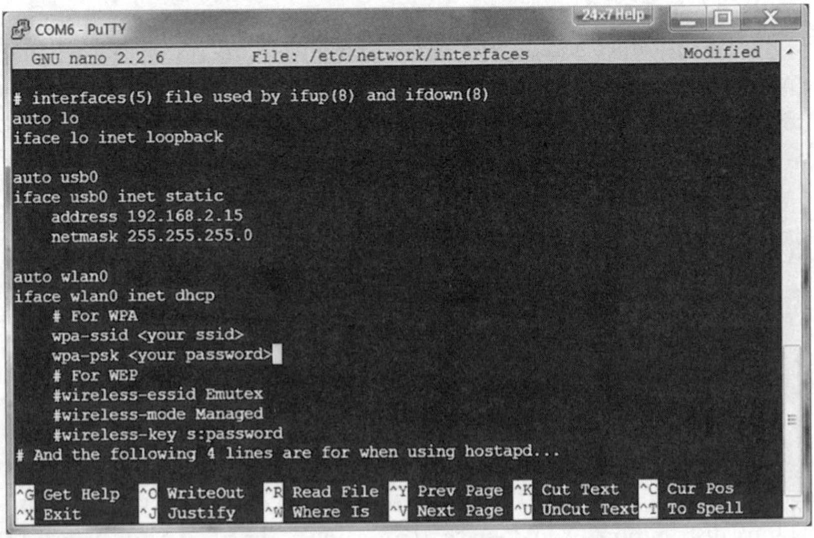

Figure 6-4 *Interfaces configuration file.*

```
root@ubilinux:/home/edison# /etc/init.d/networking restart
Running /etc/init.d/networking restart is deprecated because it may not re-enabl
e some interfaces ... (warning).
Reconfiguring network interfaces...Internet Systems Consortium DHCP Client 4.2.2
Copyright 2004-2011 Internet Systems Consortium.
All rights reserved.
For info, please visit https://www.isc.org/software/dhcp/

Listening on LPF/wlan0/fc:c2:de:3d:e6:39
Sending on   LPF/wlan0/fc:c2:de:3d:e6:39
Sending on   Socket/fallback
DHCPDISCOVER on wlan0 to 255.255.255.255 port 67 interval 7
DHCPDISCOVER on wlan0 to 255.255.255.255 port 67 interval 11
DHCPREQUEST on wlan0 to 255.255.255.255 port 67
DHCPOFFER from 192.168.1.1
DHCPACK from 192.168.1.1
bound to 192.168.1.13 -- renewal in 38630 seconds.
done.
root@ubilinux:/home/edison#
```

Figure 6-5   *Networking restart screenshot.*

DHCP protocol. Take note of this IP address, as you will need it shortly to establish a remote communications link using SSH. Figure 6-5 shows the result of the networking restart.

It will now be possible to establish a remote, wireless connection from a laptop to the Edison, once the WiFi is working and connected to your local network. This mode of operation is often referred to as headless, which, upon reflection, is the only way to control the Edison, since it has neither an organic video display output nor any direct keyboard/mouse inputs.

SSH

SSH is short for *secure shell*, which is an encrypted protocol allowing for command-line interactions between the Edison and a remote networked computer. Figure 6-6 shows the login sequence between my Macbook Pro and the Edison

```
                        donnorris — ssh — 80×24
Dons-MacBook-Pro:~ donnorris$ ssh edison@192.168.0.16
edison@192.168.0.16's password:
Linux ubilinux 3.10.17-poky-edison-ww42+ #4 SMP PREEMPT Wed Oct 29 12:41:25 GMT
2014 i686

The programs included with the Debian GNU/Linux system are free software;
the exact distribution terms for each program are described in the
individual files in /usr/share/doc/*/copyright.

Debian GNU/Linux comes with ABSOLUTELY NO WARRANTY, to the extent
permitted by applicable law.
Last login: Wed Jan 28 14:35:54 2015
edison@ubilinux:~$
```

Figure 6-6   *SSH login.*

mounted on the dev board. The Edison was connected at that time to a Windows machine using a USB cable and executing the *putty* application.

You should notice that I used the IP address assigned during the WiFi startup as part of the login. You will also have to enter the password as part of the login in exactly the same way as you did using the *putty* connection method. You are now interacting with the dev board in exactly the same way as you would if you were connected using a USB cable. I will use SSH for the rest of the chapter to demonstrate the Python concepts, which are next on the agenda.

Basic Python

The Python language application has two distinct ways of being run. The first, which I will call the *interactive mode* but others will call the *interpreted mode*, is a procedure in which a Python command, statement, or expression is executed, or run, as it is encountered. This would be the situation, as shown in Figure 6-7, in which the expression 5 * 6 is entered immediately following the Python prompt (>>>).

You can see that the expression is immediately interpreted, or evaluated, and the result displayed on the next line. This is a very convenient and useful way to interact with Python; however, it is not very practical to enter and execute anything but the most trivial of programs.

This leads to the second method of using Python: the *script mode*, also known as the *program mode*. In a script or program, which I will now refer to only as a program, all the statements or expressions are contained in a text file that is executed by Python according to strict rules enforced by the Python interpretive engine. The following is a very simple example of a Python program that I created to demonstrate this operational mode. I used the nano editor to create the following program named Counter.py, which simply displays the integers 1 through 10, as well as the square and cube of the integer.

```
●  ●  ●                    ⬆ donnorris — ssh — 80×24
root@ubilinux:/home/edison# python
Python 2.7.3 (default, Mar 14 2014, 11:57:14)
[GCC 4.7.2] on linux2
Type "help", "copyright", "credits" or "license" for more information.
>>> 5 * 6
30
>>> ▮
```

Figure 6-7   *Python expression in interactive mode.*

```
# Counter.py
# D. J. Norris 1/2015
# Code is in the public domain

# print the column headers
print('num square cube')
# loop to iterate numbers 1 through 10
for i in range(1, 11):
    print(i, i * i, i* i * i); # notice the indent
```

A few words regarding this code. Notice that the comments all start with the # symbol. The `loop` construct uses 11 as the maximum number to signal when the looping stops because the number 10 needs to be displayed. The `print` statement that calculates and displays the number, square, and cube must be indented by several spaces because this is the way you indicate to Python all the statements that belong in a defined code block. This is different from either C/C++ or Node.js, both of which use { } brackets to show a defined block of code. Just remember, white space does matter when it comes to writing Python programs.

You run the code by entering the following, assuming you are running Python from the `root` level:

```
python Counter.py
```

Figure 6-8 shows the result of executing the above command line.

Python also has a very comprehensive help feature, which I will explore in the next section.

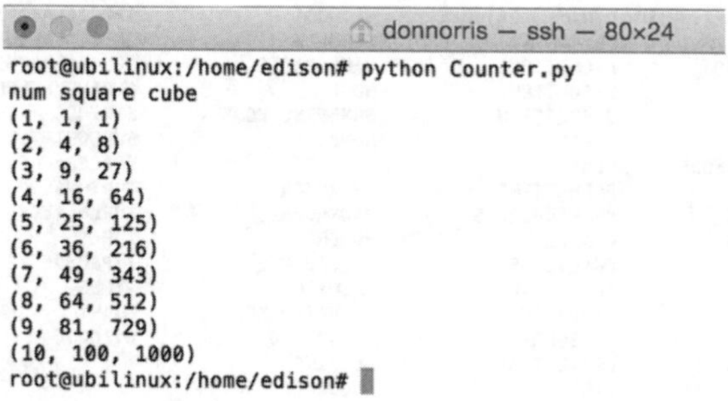

```
●  ●  ●                 ⌂ donnorris — ssh — 80×24
root@ubilinux:/home/edison# python Counter.py
num square cube
(1, 1, 1)
(2, 4, 8)
(3, 9, 27)
(4, 16, 64)
(5, 25, 125)
(6, 36, 216)
(7, 49, 343)
(8, 64, 512)
(9, 81, 729)
(10, 100, 1000)
root@ubilinux:/home/edison# ▌
```

Figure 6-8   *Counter.py program results.*

Python Help

The Python help feature is activated by entering help() at the Python interactive prompt, as shown in Figure 6-9. Also in the figure, you can see the results of my entering keywords and then topics. Thirty-one keywords were listed as

```
root@ubilinux:/home/edison# python
Python 2.7.3 (default, Mar 14 2014, 11:57:14)
[GCC 4.7.2] on linux2
Type "help", "copyright", "credits" or "license" for more information.
>>> help()

Welcome to Python 2.7! This is the online help utility.

If this is your first time using Python, you should definitely check out
the tutorial on the Internet at http://docs.python.org/2.7/tutorial/.

Enter the name of any module, keyword, or topic to get help on writing
Python programs and using Python modules.  To quit this help utility and
return to the interpreter, just type "quit".

To get a list of available modules, keywords, or topics, type "modules",
"keywords", or "topics".  Each module also comes with a one-line summary
of what it does; to list the modules whose summaries contain a given word
such as "spam", type "modules spam".

help> keywords

Here is a list of the Python keywords.  Enter any keyword to get more help.

and                 elif                if                  print
as                  else                import              raise
assert              except              in                  return
break               exec                is                  try
class               finally             lambda              while
continue            for                 not                 with
def                 from                or                  yield
del                 global              pass

help> topics

Here is a list of available topics.  Enter any topic name to get more help.

ASSERTION           DEBUGGING           LITERALS            SEQUENCEMETHODS2
ASSIGNMENT          DELETION            LOOPING             SEQUENCES
ATTRIBUTEMETHODS    DICTIONARIES        MAPPINGMETHODS      SHIFTING
ATTRIBUTES          DICTIONARYLITERALS  MAPPINGS            SLICINGS
AUGMENTEDASSIGNMENT DYNAMICFEATURES     METHODS             SPECIALATTRIBUTES
BACKQUOTES          ELLIPSIS            MODULES             SPECIALIDENTIFIERS
BASICMETHODS        EXCEPTIONS          NAMESPACES          SPECIALMETHODS
BINARY              EXECUTION           NONE                STRINGMETHODS
BITWISE             EXPRESSIONS         NUMBERMETHODS       STRINGS
BOOLEAN             FILES               NUMBERS             SUBSCRIPTS
CALLABLEMETHODS     FLOAT               OBJECTS             TRACEBACKS
CALLS               FORMATTING          OPERATORS           TRUTHVALUE
CLASSES             FRAMEOBJECTS        PACKAGES            TUPLELITERALS
CODEOBJECTS         FRAMES              POWER               TUPLES
COERCIONS           FUNCTIONS           PRECEDENCE          TYPEOBJECTS
COMPARISON          IDENTIFIERS         PRINTING            TYPES
COMPLEX             IMPORTING           PRIVATENAMES        UNARY
CONDITIONAL         INTEGER             RETURNING           UNICODE
CONTEXTMANAGERS     LISTLITERALS        SCOPING
CONVERSIONS         LISTS               SEQUENCEMETHODS1

help>
```

Figure 6-9 *Python help screen.*

well as 79 topics. I selected the keyword `for` to show more detailed help, which is shown in Figure 6-10.

An example of a help screen for the topic `INTEGER` is shown in Figure 6-11. Entering the word `modules` in the help system produced 238 entries, as shown in Figure 6-12.

Each `module` help entry averages several page screens in length. I chose not to show a typical `module` help entry because it is somewhat confusing to a beginning Python developer. Just be aware that these impressive resources are readily available in much the same way that the APIs are available on the Web for other languages, such as C and C++.

```
help> for
The ``for`` statement
*********************

The ``for`` statement is used to iterate over the elements of a
sequence (such as a string, tuple or list) or other iterable object:

    for_stmt ::= "for" target_list "in" expression_list ":" suite
                 ["else" ":" suite]

The expression list is evaluated once; it should yield an iterable
object.  An iterator is created for the result of the
``expression_list``.  The suite is then executed once for each item
provided by the iterator, in the order of ascending indices.  Each
item in turn is assigned to the target list using the standard rules
for assignments, and then the suite is executed.  When the items are
exhausted (which is immediately when the sequence is empty), the suite
in the ``else`` clause, if present, is executed, and the loop
terminates.

A ``break`` statement executed in the first suite terminates the loop
without executing the ``else`` clause's suite.  A ``continue``
statement executed in the first suite skips the rest of the suite and
continues with the next item, or with the ``else`` clause if there was
no next item.

The suite may assign to the variable(s) in the target list; this does
not affect the next item assigned to it.

The target list is not deleted when the loop is finished, but if the
sequence is empty, it will not have been assigned to at all by the
loop.  Hint: the built-in function ``range()`` returns a sequence of
integers suitable to emulate the effect of Pascal's ``for i := a to b
do``; e.g., ``range(3)`` returns the list ``[0, 1, 2]``.

Note: There is a subtlety when the sequence is being modified by the loop
   (this can only occur for mutable sequences, i.e. lists). An internal
   counter is used to keep track of which item is used next, and this
   is incremented on each iteration.  When this counter has reached the
   length of the sequence the loop terminates.  This means that if the
   suite deletes the current (or a previous) item from the sequence,
   the next item will be skipped (since it gets the index of the
   current item which has already been treated).  Likewise, if the
   suite inserts an item in the sequence before the current item, the
   current item will be treated again the next time through the loop.
   This can lead to nasty bugs that can be avoided by making a
   temporary copy using a slice of the whole sequence, e.g.,

      for x in a[:]:
          if x < 0: a.remove(x)

Related help topics: break, continue, while

help>
```

Figure 6-10 *for keyword help screen.*

```
help> INTEGER
Integer and long integer literals
**********************************

Integer and long integer literals are described by the following
lexical definitions:

    longinteger    ::= integer ("l" | "L")
    integer        ::= decimalinteger | octinteger | hexinteger | bininteger
    decimalinteger ::= nonzerodigit digit* | "0"
    octinteger     ::= "0" ("o" | "O") octdigit+ | "0" octdigit+
    hexinteger     ::= "0" ("x" | "X") hexdigit+
    bininteger     ::= "0" ("b" | "B") bindigit+
    nonzerodigit   ::= "1"..."9"
    octdigit       ::= "0"..."7"
    bindigit       ::= "0" | "1"
    hexdigit       ::= digit | "a"..."f" | "A"..."F"

Although both lower case ```l``` and upper case ```L``` are allowed as
suffix for long integers, it is strongly recommended to always use
```L```, since the letter ```l``` looks too much like the digit
```1```.

Plain integer literals that are above the largest representable plain
integer (e.g., 2147483647 when using 32-bit arithmetic) are accepted
as if they were long integers instead. [1] There is no limit for long
integer literals apart from what can be stored in available memory.

Some examples of plain integer literals (first row) and long integer
literals (second and third rows):

    7     2147483647                      0177
    3L    79228162514264337593543950336L  0377L   0x100000000L
          79228162514264337593543950336           0xdeadbeef

Related help topics: int, range

help>
```

Figure 6-11   *INTEGER help topic.*

You exit the Python help system by pressing the enter key while the help
prompt is shown.

I strongly encourage you to take advantage of this comprehensive help system. While it will not replace a good textbook on Python programming, it will
provide some immediate help in resolving issues that will pop up while you are
programming.

Now on to some additional basic Python programming concepts.

Data Types, Variables, and Constants

I first introduced various data types in Chapter 3 during the Processing language
discussion. Processing is based on the C/C++ languages and is strongly typed as
mentioned in that chapter. Python is implicitly typed and, therefore, does not
require variables to be explicitly declared as particular types before being used.
In my opinion, this might cause some problems for beginners because it does not

```
Tkinter            colorsys          mimetypes         sunau
UserDict           commands          mimify            sunaudio
UserList           compileall        mmap              symbol
UserString         compiler          modulefinder      symtable
_LWPCookieJar      contextlib        multifile         sys
_MozillaCookieJar  cookielib         multiprocessing   sysconfig
__builtin__        copy              mutex             syslog
__future__         copy_reg          netrc             tabnanny
_abcoll            crypt             new               tarfile
_ast               csv               nis               telnetlib
_bisect            ctypes            nntplib           tempfile
_bsddb             curses            ntpath            termios
_codecs            datetime          nturl2path        test
_codecs_cn         dbhash            numbers           textwrap
_codecs_hk         dbm               opcode            this
_codecs_iso2022    debconf           operator          thread
_codecs_jp         decimal           optparse          threading
_codecs_kr         difflib           os                time
_codecs_tw         dircache          os2emxpath        timeit
_collections       dis               ossaudiodev       tkColorChooser
_csv               distutils         parser            tkCommonDialog
_ctypes            dl                pdb               tkFileDialog
_ctypes_test       doctest           pickle            tkFont
_curses            dumbdbm           pickletools       tkMessageBox
_curses_panel      dummy_thread      pipes             tkSimpleDialog
_elementtree       dummy_threading   pkgutil           toaiff
_functools         email             platform          token
_hashlib           encodings         plistlib          tokenize
_heapq             errno             popen2            trace
_hotshot           exceptions        poplib            traceback
_io                fcntl             posix             ttk
_json              filecmp           posixfile         tty
_locale            fileinput         posixpath         turtle
_lsprof            fnmatch           pprint            types
_multibytecodec    formatter         profile           unicodedata
_multiprocessing   fpectl            pstats            unittest
_pyio              fpformat          pty               urllib
_random            fractions         pwd               urllib2
_socket            ftplib            py_compile        urlparse
_sqlite3           functools         pyclbr            user
_sre               future_builtins   pydoc             uu
_ssl               gc                pydoc_data        uuid
_strptime          genericpath       pyexpat           warnings
_struct            getopt            quopri            wave
_symtable          getpass           random            weakref
_sysconfigdata     gettext           re                webbrowser
_sysconfigdata_nd  glob              readline          whichdb
_testcapi          grp               repr              wsgiref
_threading_local   gzip              resource          xdrlib
_warnings          hashlib           rexec             xml
_weakref           heapq             rfc822            xmllib
_weakrefset        hmac              rlcompleter       xmlrpclib
abc                hotshot           robotparser       xxsubtype
aifc               htmlentitydefs    runpy             zipfile
antigravity        htmllib           sched             zipimport
anydbm             httplib           select            zlib
argparse           ihooks            sets
array              imageop           sgmllib

Enter any module name to get more help.  Or, type "modules spam" to search
for modules whose descriptions contain the word "spam".

help>
```

Figure 6-12 *modules* help listing.

enforce a strict discipline regarding how data is assigned and/or converted between different types. The following interactive session shown in Figure 6-13 illustrates this problem.

```
● ● ●
>>> h = 5
>>> h / 2
2
>>> h = 5.0
>>> h / 2
2.5
>>> ▮
```

Figure 6-13 *Potential problem with variable data types.*

In the first case, the *h* variable is assigned the integer value 5, thus making it an integer variable. Dividing *h* by 2 generates the correct result of 2, as *h* can only be a whole number. This result will likely cause a beginner some head scratching, as a dividend of 2.5 might have been expected, however erroneous. The second part of the example shows *h* being reassigned with the real number value of 5.0. This time, the division by 2 results in a 2.5 value: what a beginner would naturally expect. The variable *h* is now a "real typed" variable and can hold numbers with a decimal point. The previous example is just one small issue that Python programmers must keep in mind.

Sometimes you will need to use a specific value repeatedly, such as the value for Pi, in arithmetic operations. Python versions 2.7 or 3.0 really do not allow for constants to be created. It is suggested that you simply create a statement in the beginning of your program, such as PI = 3.14159265, and simply use PI whenever you need the constant. Just don't reassign it later in the program by doing something like PI = 3.0. It is your responsibility to ensure that constants remain just that, constant.

String constants, also known as string literals, are also very handy to use in a program. They can reduce the coding effort substantially if the same text data is repeatedly used, such as an address. Figure 6-14 shows an example

Another way to improve program efficiency and readability is to use functions, or methods as I referred to them in earlier chapters. In Python, the term *function* is used to represent compact units of code that usually perform one or two tasks. They are called by a name with or without arguments or parameters, as suits the

```
● ● ●                      ⌂ donnorris — ssh — 80×24
>>> address = '322 Casa Lane, Pasadena, CA 99672'
>>> address
'322 Casa Lane, Pasadena, CA 99672'
>>> ▮
```

Figure 6-14 *String literal example.*

function design. Python already contains many prebuilt or native functions, such as `print`. The ones I will discuss next are user-defined functions.

User-Defined Functions

I believe a good way to demonstrate how to define and use a Python function is by presenting a practical example. For this example, I intend to interface a temperature sensor with a specific calibration chart to the dev board. The calibration chart shows the sensor's voltage output versus temperature input. Figure 6-15 is an example of this sensor's calibration chart.

The line on the chart is also called its transfer curve, which, in this example, was also created to match a predetermined mathematical equation. I used the following equation to make the function programming easier:

$$temperature = K_p * (1 - e^{-volts})$$

where K_p is a constant of proportionality and
volts is the input in the range of 0 to 3.3 V

The conversion process of changing a raw count to an equivalent temperature is a task that would be repeatedly called in a real-world situation, and thus, would be an ideal case for a user-defined function.

This function, which converts a raw ADC count to a final temperature in F°, is named `raw_to_temp( <integer> )`, where the word `integer` shows that it takes a whole number as an argument. The function returns a float or real

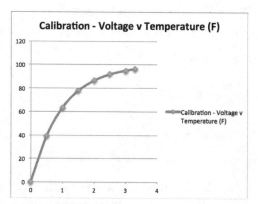

Figure 6-15 *Temperature sensor calibration curve.*

number, which is the desired temperature. The `integer` argument comes from the ADC and ranges from 0 to 1023 because the Edison has a 10-bit ADC. In this example, an output of 1023 would represent a maximum input of 3.3 V, while a 0 would naturally represent a 0 V input.

The following Python code listing is named VoltToTemp.py and contains the user-defined function as well as some test code to display a series of temperatures reflecting sensor voltages from 0 to 3 V in 0.5-V increments. I have added many comments to the listing to help clarify what is happening at any point in the program. Figure 6-16 shows the result of running this program.

```
# VoltToTemp.py
# D. J. Norris 1/2015
# Code is in the public domain

# require the math library for the exp() function
import math

# need to define a user function before using it in a Python program.
# The keyword def is required
def raw_to_temp(inData):
  # notice the float cast before inData. You need it; otherwise it
    results in a 0 dividend.
  # see my discussion about this issue in the chapter text
  volt = (float(inData)/1024)*3.30
  # this is where the voltage to temp conversion takes place
  temp = 100*(1 - (math.exp(-volt)))
  # now return the converted value back to the calling point
  return temp

# initialize the data 'transfer' variable
raw = 0.00

# print the column headers before the loop starts
print('voltage temp')

# the odd step value of 155 represents 0.5 volts for a 10 bit ADC with
  a 3.3-V max input
for i in range(0,1024,155):
  raw = i
  temp = raw_to_temp(raw)
  # convert the ADC count back to voltage. Makes it easier to read
    the data
  volt = (float(raw)/1024)*3.30
  # round the data to no more than two decimal points. The default
    is 10
  print(round(volt,2), round(temp,2))
```

As you can see, the displayed temperatures match very precisely with the calibration chart due to the precise mathematical equation that was used to

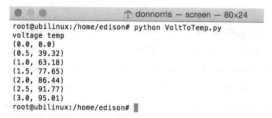

Figure 6-16 *VoltToTemp.py program results.*

create both the chart and the program results. While a mathematical equation is great to ease the calibration process, in reality, such a match does not often occur. What does happen is an empirical calibration curve is supplied containing a series of data points generated by the manufacturer. It would then be up to the sensor's user to interpolate, as necessary, for points not exactly matching the manufacturer's data. This is the topic for the next section in which I demonstrate how to use interpolation with a calibration data set to obtain the true sensor readings.

Interpolated Sensor Measurements

I have modified the user-defined function to accept the data shown in Table 6-1, which is representative of a manufacturer's calibration data set for the temperature sensor in the example.

Voltage	Temperature
0	0
0.5	38.4
1.0	65.2
1.5	76.3
2.0	87.6
2.5	94.9
3.0	97.5

Table 6-1 *Sensor Voltage/Temperature Calibration*

The data set must be present in a file in the same directory as the modified VoltsToTemp.py program, which I renamed to VoltToTemp2.py. I also named the calibration file Cal.dat to reflect its content.

I used the two list objects in this program to hold the voltage and temperature data that are read into the program as a *comma separated variable* (csv) file. csv files are also commonly used as a file store for Excel data. Lists are used extensively in Python programs, which is the reason that I used them in this program.

```
# VoltToTemp2.py
# D. J. Norris 1/2015
# Code is in the public domain

# need to use the csv library to read the comma separated variable
# file (csv) 'cal.dat'
import csv

# open the cal file and use a local name 'f' with open('cal.dat')
  as f:

  # need a reader object to gain access to the cal data
  reader = csv.reader(f)

  # two lists to hold the voltage and temperature cal data
  volt = list()
  temp = list()

  # read all the rows in; put the data into the lists for row in
    reader:
     volt.append(row[0])
     temp.append(row[1])

# this is the interpolation function. It takes an ADC count argument
def interpolate(inData):
  max = 0
  # cast the inData to float or you will get 0s
  v = (float(inData)/1024)*3.30

  for x in range(2, len(volt)):
    if v >= float(volt[x]):
      max = x + 1
      break

  # both the volt and temp lists must be cast to float from string
  # classic interpolation data crunching done next
  v1 = v - float(volt[max-1])
  v2 = float(volt[max]) - float(volt[max-1])
  t2 = float(temp[max]) - float(temp[max-1])
  t1 = (t2 * v1)/v2 + float(temp[max-1])
  # return the interpolated temperature
  return t1
```

Figure 6-17   *VoltToTemp2.py program results.*

```
print('voltage  temp')

volts = 0.0

# mixed up the voltage input to demonstrate interpolated data
for i in range(120,980,140):
  temp1 = interpolate(i)
  volts = (float(i)/1024)*3.3
  # rounded the results for easier viewing
  print(round(volts,2), round(temp1,2))
```

Figure 6-17 is the result of running the VoltToTemp2.py program. It appears the interpolation routine worked as expected, based on the displayed results.

This last program brings this chapter to an end; however, there is much more information regarding the Python language, its libraries, and its interfacing techniques to be discussed in the following chapter.

Summary

Some essential Python concepts were explained and explored through practical examples to provide you with a reasonable background to start creating your own Python programs for the Edison. The next few chapters will more fully develop these introductory ideas and, hopefully, provide a good basis for your future development efforts.

7

Python Classes, Methods, and the libmraa Library

This chapter could have been subtitled Python and object-orientation (OO), since I include a discussion of Python objects and classes from an OO perspective. I will start the OO discussion with a broad overview.

Basic OO Concepts

OO in computer language development has been in existence for a relatively long time (in computer terms) dating from the introduction of Simula in the 1960s. It is based upon four principles:

1. **A**bstraction
2. **P**olymorphism
3. **I**nheritance
4. **E**ncapsulation

I intentionally bolded the first letters of each principle to form an acronym, APIE, or Apple PIE, which should help you remember these key elements of any OO language. I do wish to state that an OO language is not necessarily better or worse than non-OO languages, such as C. It really depends upon the project requirements that are to be met as to which language is most suitable. In fact, it wasn't until fairly recently that embedded developers even had the opportunity to use an OO language. In the early days of embedded development, which would have been 1970s to mid-1990s, C or assembly language was pretty much all that was available. It was the introduction of sophisticated 32-bit microprocessor cores along with much larger onboard memory that made OO possible in the embedded realm

I will start this OO discussion by focusing on the *class*, which is really the critical element in any OO language.

The Class

Being able to logically represent a real-world item is the sole purpose of a class. Embedded systems are truly about being able to interact with real things, such as sensors, actuators, etc. The class is the logical representation of things we need to use and, as such, contains both descriptors and behaviors about these things. In OO parlance, descriptors are known as *attributes*, and behaviors as *methods*, a language construct we have already encountered in earlier chapters.

I will first create a class modeling an electronic motor controller (EMC), as it is a real-world item used quite frequently in Edison projects. Electric motors are relatively simple devices that typically have a single rotating part, normally called a rotor. However, motors often require higher voltages and currents than a microcontroller can directly and safely provide. This is the reason why the EMC is used as an interface between a microcontroller, such as the Edison, and an electric motor.

After reflecting a moment on what constitutes a real EMC, you would soon realize that only a few attributes are needed, and likewise, only a few methods are required to adequately model a controller for our OO purposes. I have listed some attributes and methods in Table 7-1 that seemed to me to adequately model a simple EMC.

Name	Attribute	Method	Description
Voltage	x		Maximum input voltage
Current	x		Maximum input current
r/min	x		Maximum r/min
on( )		x	Turn on motor controller
off( )		x	Turn off motor controller
moveFwd(n)		x	Move forward, n proportional to speed
moveBack(n)		x	Move back, n proportional to speed

Table 7-1 *Electronic Motor Controller Attributes and Methods*

Often, the `on()` and `off()` methods are optional, as the EMC is on all the time, but they will energize the motor when move commands are received.

EMC design is dependent upon the type of motor being controlled. I tend to use three types of motors in my projects. These are:

1. *Servos*—restricted range of motion normally $+/-90°$ for a total of $180°$

2. *Continuous rotation servos*—continuous rotation either CW or CCW

3. *Analog motors*—these may be brushed or unbrushed whose rotational speed and direction are based upon applied voltage and current

The manner in which an EMC handles a `moveForward()` or `moveBack()` command will be directly related to the type of motor it is designed to control. This statement leads me back to an OO design that I created for this EMC hierarchy. This is shown in Figure 7-1 in a Unified Modeling Language (UML) diagram, which is a very common way of illustrating OO class relationships.

The class labeled EMC is known as a parent class because it contains both attributes and methods common to all three classes depicted below the EMC class. These classes are named:

1. Servo

2. CRServo

3. AnalogMotor

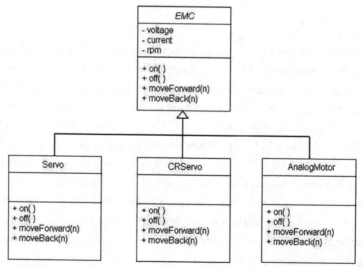

Figure 7-1 *EMC class diagram.*

These classes are collectively known as *child classes* because they inherit all the attributes and methods of the parent class. However, since they implement moveForward( ) and moveBack( ) in different ways, they are more specific than is possible with the EMC class. I like to consider the EMC class as more of an abstract class in which it does not make sense to instantiate (create) a real-world object from its general definition. It is only from the child classes that useful objects may be instantiated to perform actual operations in a Python program, as I will demonstrate in the code snippet below. This class diagram is a classic demonstration of inheritance, one of the key OO principles.

Each method in the child classes will be implemented in a different manner because it is based upon a different hardware configuration. Take for example, the CRServo class and its moveForward( ) method. Recall in Chapter 4, I used CR servos to drive the autonomous robot car and used a program named Servo2 written in the Processing language to control the car. The following is a snippet from that program:

```
void loop()
{
    myservo1.write(maxSpeedBack);
    myservo2.write(maxSpeedFwd);
// check for an obstacle
obstacle = digitalRead(signal);
  if(obstacle) {
      avoid(); // call the obstacle avoidance method
  }
}
```

The statement myservo1.write(maxSpeedBack) is an example of the class method named write being invoked by an object named myservo1, which, in turn, was instantiated from a class named Servo. The method argument, maxSpeedBack, is just an integer number representing full speed servo rotation in a particular direction (CW or CCW). The moveForward method in the CRServo class would be equivalent to the Servo class write method with a specific numerical range. The moveBack method would also be equivalent to the write method with a different numerical range designed to rotate the CR servo in the opposite direction as compared to the moveForward method.

The other major consideration for method implementations is the way in which the microcontroller interfaces to the EMC. There are at least two approaches to creating this interface.

1. Use a serial communications protocol, such as I2C or SPI, with commands sent to the EMC as a series of numbers and characters. This interface type is highly flexible and also provides a relatively easy means of controlling a motor or servo.

2. Use a general-purpose input/output (GPIO) line to create digital pulses, which are then directly input into the servo. This is exactly how the CR servos were controlled in the robot car.

The `write` method used in the second approach toggles the GPIO pin "attached" to a servo by this Processing statement:

```
myservo1.attach(3);   // attaches a servo on pin 3 to the
myservo1 object
```

At this point in the OO discussion, I have to deviate a bit and show you how to install the libmraa library. The availability of this library is crucial because it provides the Python classes and methods to implement key methods, such as a `write` method, which is equivalent to the one in the Processing library. The libmraa library contains all the low-level drivers and associated logic required to be able to create the classes that allow creating Python hardware interface classes.

Installing the libmraa Library

The libmraa library is an open-source development effort created using C/C++ but also containing bindings for use with Python, Javascript, and Node.js. Port names and pin numbering have been tailored to match, as far as possible, the actual hardware that will use the library. As of version 0.5.4, libmraa works very well with the Intel Galileo and Edison platforms; however, other platforms can easily be used, as the library incorporates dynamic board detection. This means that it is possible for low-level communications between a processor and a sensor or actuator to be established at runtime without needing a tailored kernel compilation.

It turned out that there were no precompiled Python libmraa versions available when this book was written, meaning that I could not simply use the aptitude repository manager (apt-get) application to install libmraa from a Debian repository. This meant that the library would have to be compiled from source code and separately installed. That process is not too hard, thanks to a great tutorial named "Installing libmraa on Ubilinux for Edison" available from learn .sparkfun.com. Many thanks to Casey D for his fine effort. I have condensed this tutorial to a series of steps that you should carefully follow to ensure that a working copy of libmraa installs in the Debian distribution.

NOTE   *Ensure you are at the* root *level before starting this procedure.*

1.	Update the Debian distribution	`apt-get update`
2.	Install pcre	`apt-get install libpcre3-dev`
3.	Install git	`apt-get install git`
4.	Install cmake	`apt-get install cmake`
5.	Install python development	`apt-get install python-dev`
6.	Install swig	`apt-get install swig`
7.	Git mraa	`git clone https://github.com/intel-iot-devkit/ mraa.git`
8.	Create the build directory & cd	`mkdir mraa/build && cd $_`
9.	Compile mraa with a flag	`cmake .. -DBUILDSWIGNODE=OFF`
10.	Make the compilation	`make`
11.	Install it	`make install`
12.	Change directory	`cd`

At this point, you will have compiled and installed the mraa library, but you will not yet be able to use it with Python until you create a link. Before I show you how that is done, I want to refer you back to the original tutorial on the learn .sparkfun.com website, which shows you how to use mraa with C/C++, in case you desire to go that path.

Using mraa with Python requires that the mraa location and path be exported so that Python knows where to find the library. The command to do this is:

```
export PYTHONPATH=$PYTHONPATH:$(dirname $(find /usr/
local -name mraa.py))
```

Unfortunately, you will have to type the above command every time you want to use mraa with Python unless you follow the next procedure, which permanently puts this export command in the bash configuration file.

1. Start nano for the bash configuration file.

   ```
   nano ~/.bashrc
   ```

2. Scroll to the bottom and add the export command shown above.

3. Save and exit the editor.

4. Reboot to make the change effective.

If you use the `sudo` command, you should refer to the tutorial on the learn
.sparkfun.com website regarding some more changes that might be helpful.

mraa Version Check

The last step in ensuring that the mraa installation works successfully with
Python is creating an extremely simple program whose only function is to display the mraa version number. Because the program is so short, I simply entered
it line-by-line in an interactive session. This session is shown in Figure 7-2.

Blink Program

I next created a blink program to check out the mraa GPIO interface. I used pin
13 on the dev board, since there is a permanent LED attached to that pin. The
blink program I used is named mraaBlink.py and is listed below:

```
# mraaBlink.py
# D. J. Norris 2/2015
# Code is in the public domain

# need to import both mraa and time libraries
import mraa
import time

# useful constants
on = 1
```

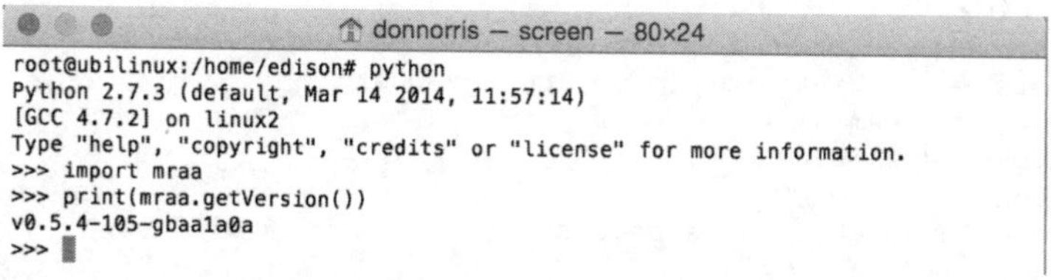

```
root@ubilinux:/home/edison# python
Python 2.7.3 (default, Mar 14 2014, 11:57:14)
[GCC 4.7.2] on linux2
Type "help", "copyright", "credits" or "license" for more information.
>>> import mraa
>>> print(mraa.getVersion())
v0.5.4-105-gbaa1a0a
>>>
```

Figure 7-2 *Interactive session displaying the mraa version number.*

```
off = 0

# initialization
led = mraa.Gpio(13)
led.dir(mraa.DIR_OUT)

# forever loop
while True:
    led.write(on)
    time.sleep(1)
    led.write(off)
    time.sleep(1)
```

You run the program by entering:

```
python mraaBlink.py
```

The LED on the dev board blinked on and off at one-second intervals for as long as the program was active. The program immediately stopped when I entered a Control-C (^C) on the laptop keyboard that was connected to the dev board. This action is in sharp contrast to the Intel Arduino IDE in which the LED would continue to blink until a new program was loaded into the board.

Servo Control Program

This next program uses one of the *pulse-width modulation* (PWM) outputs available on the dev board. In this case, I used pin 3, but there are several others available, if needed. PWM is very useful in controlling servos and regular electric motors. I created the following program to generate a PWM waveform that repeatedly cycles from a 0% pulse duration to a 100% pulse duration which, if connected to a typical analog servo, would cause it to oscillate throughout its full 180° range of motion. The program is named PwmTest.py and is listed below.

```
# PwmTest.py
# D. J. Norris 2/2015
# Code is in the public domain

#need both mraa and time libraries
import mraa
import time

# Use pin 3; it is a pwm pin on the dev board
pwmPin = mraa.Pwm(3)
pwmPin.period_us(2000)
pwmPin.enable(True)
# pulse HIGH duration as a percentage of period = 2000 µsec
tau = 0.0
```

```
# forever loop
while True:
    pwmPin.write(tau)
    time.sleep(0.05)  # 50-msec delay
    tau = tau + 0.01
    if tau >= 1:  # check on tau's value; do not exceed 1.0
        tau = 0.0  # reset to 0
```

I connected a mini servo to the dev board to test how the program operated it. Also note that I used a separate five pack of AA batteries to power the servo. I would not recommend using the dev board 5-V power supply because the servo will take more current than is available from this source. When that happens, the dev board will begin to function erratically and may even shutdown. Figure 7-3 is a picture of the setup.

The program was run by entering:

```
python PwmTest.py
```

The servo started oscillating in its full range of motion of 180° when the enter key was pressed after the above command was entered.

Figure 7-4 is a snapshot of the waveform generated by the program. This figure was taken from a screenshot of a model 3406B Pico USB oscilloscope monitoring pin 3 on the dev board.

The pulse waveform shown in the figure expands its high time from the left to the right at a frequency of 0.500 kHz, which is equivalent to the 2000-μsec period specified by the statement `pwmPin.period_us(2000)`.

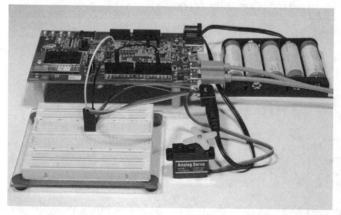

Figure 7-3 *Mini servo test setup.*

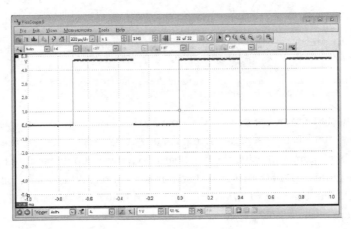

Figure 7-4 *Snapshot of PwmTest program output.*

I next modified the PwmTest program to control a CR servo. This modified program is discussed in the next section.

CR Servo Control Program

CR servos are easily controlled by using the Python mraa library. The following program is named CRServoTest and is a slightly modified version of the PwmTest program.

```
# CRServoTest.py
# D. J. Norris 2/2015
# Code is public domain

#need both mraa and time libraries
import mraa
import time

# Use pin 3; it is a pwm pin on the dev board
pwmPin = mraa.Pwm(3)
pwmPin.period_us(2000)
pwmPin.enable(True)
# pulse HIGH duration as a percentage of period = 2000 µsec
tau = 0.1

# forever loop
while True:
  pwmPin.write(tau)
  time.sleep(0.05) # 50-msec delay
```

I connected a CR servo to the dev board to test how the program operated it. Figure 7-5 is a picture of the setup. The CR servo I used was one of the units that

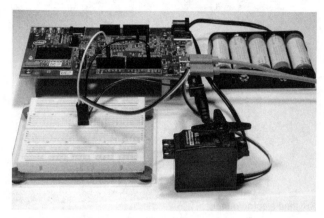

Figure 7-5 *CR servo test setup.*

powered the robot car in Chapter 4. It also is connected to a separate 7.5-V battery pack, as was done for the mini servo test. A single GPIO line connected at pin 3 controls the CR servo's speed and rotational direction.

The program was run by entering:

```
python CRServoTest.py
```

The servo started rotating in the CW direction at approximately 40 r/min when the enter key was pressed after the above command was entered.

I also connected the USB oscilloscope to pin 3 in order to observe the PWM waveform. Figure 7-6 shows the pulse waveform output for this program. It is

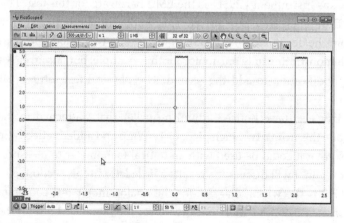

Figure 7-6 *CRServoTest program pulse waveform.*

Figure 7-7   *Brushed electric motor.*

easy to see that the pulse duration is 200 μsec, which results from a multiplication of the tau value of 0.1 by the 2000-μsec period..

The last EMC to be covered is the one that controls a brushed electric motor, which is shown in Figure 7-7. The motor pictured in the figure is often referred to as a toy electric motor because that style is often used in toy products with a gear train, which reduces the r/min from very high to low, while correspondingly increasing the torque. I will use the motor alone without any gearing, as I only want to demonstrate simple motor control.

Analog Motor Control Program

I used a Texas Instrument L293D as the EMC hardware for this test. The L293D is a dual Darlington transistor driver. This chip will take both PWM and TTL inputs from the dev board and directly control up to two analog electric motors. I will be demonstrating only one motor for this example, but it is an easy control-program modification to add the second. The physical test setup is shown in Figure 7-8.

You might also have noticed a mini propeller that I attached to the motor shaft. I made this from a two-inch piece of Ty-Wrap and added reflective paper to one side so that it could be used with an optical tachometer to measure the no-load shaft r/min. I will discuss the results a little later in this section.

Figure 7-9 is the wiring diagram for the analog motor test configuration. It uses a separate 7.5-V AA battery pack to power the motor, as I did for the servos in the last two tests.

Figure 7-8   *Analog motor test setup.*

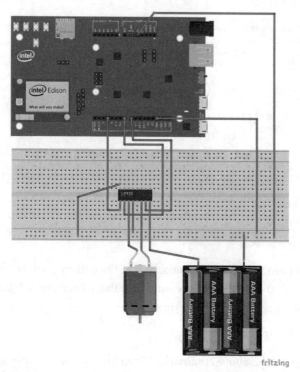

Figure 7-9   *Analog motor test wiring diagram.*

The following is the mraa-based control program for this test. It is heavily commented to help you understand what is happening in the code. The program is named AnalogMotorTest.

```
# AnalogMotorTest.py
# D. J. Norris 2/2015
# Code is in the public domain

#need both mraa and time libraries
import mraa
import time

#useful constants
on = 1
off = 0

# set up the logic control pins
in1 = mraa.Gpio(8)
in2 = mraa.Gpio(9)

# now set them as outputs
in1.dir(mraa.DIR_OUT)
in2.dir(mraa.DIR_OUT)

# now set the correct logic states for CCW rotation
# If you want CW rotation, switch (in1 off & in2 on)
in1.write(on)
in2.write(off)

# Use pin 3; it is a pwm pin on the dev board
pwmPin = mraa.Pwm(3)
pwmPin.period_us(2000) # use a 2000 µsec period
pwmPin.enable(True)

# pulse HIGH duration as a percentage of period fixed at 2000 µsec
tau = 0.5    # sets the speed somewhere near the middle of the motor's
             operating range

# forever loop
while True:
    pwmPin.write(tau)
    time.sleep(0.05) # 50-msec delay
```

Prior to running the program, ensure that the battery pack is connected, and double check all the wiring between the breadboard and the dev board. Run the program at the root level by entering:

```
python AnalogMotorTest.py
```

After entering the above command, I immediately saw the motor rotating at a moderate r/min, as compared to my initial operational check when I simply connected the battery pack directly to the motor. The motor speed is somewhat proportional to the applied voltage. Notice, I stated it was *somewhat*, which implies a non-linear relationship. A linear relationship would mean if 3 V creates 3000 r/min than 6 V should rotate it at 6000 r/min. That's not what happens in reality due to a variety of factors mostly related to how the motor is constructed.

I was interested in determining the speed versus applied-voltage relationship so I conducted the following experiment. I edited the tau value from 0.2 to 1.0 in steps of 0.2. I then reran the program at each tau step value and used an optical tachometer to measure the rotational speed of the miniature propeller attached to the motor shaft. The optical tachometer I used is shown in Figure 7-10 for those readers interested in duplicating this test.

I should also discuss why tau, which is the PWM pulse on, or HIGH time, is related to the applied voltage. A PWM pulse train applied to the motor essentially is averaged so that the real, or effective, voltage applied to the motor is directly proportional to the length of the pulse in time. This means a PWM pulse train with a tau of .5 and a 7.5-V peak voltage will effectively be half or 3.75 V. It is not critical to go into the reasons why a motor averages the pulse train waveform other than to state that a motor is a real physical device with inertia and inductive windings..

The results of using different tau values is shown in Figure 7-11. You can clearly see that tau is not linearly proportional to r/min because the curve is convex instead of being a straight line.

Figure 7-10 *Optical tachometer.*

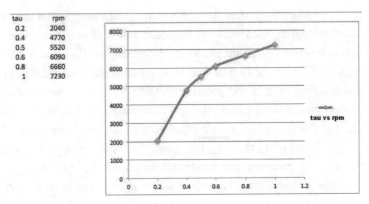

Figure 7-11 *tau versus r/min chart.*

You should also note that tau values below 0.2 do not create sufficient voltage to turn the motor rotor. In fact, I even had to flick the mini propeller at the 0.2-tau level to start the shaft rotating.

Another factor to be considered when using PWM with analog motors is the torque produced with varying voltage. I did not have the means to measure torque; however, I believe that the effective torque diminishes quite rapidly with low tau values.

All of the above results lead me to believe that analog electric motors can be effectively speed controlled with tau values from 0.4 to 1.0. Trying to go below 0.4 tau is probably not a good idea, especially when using toy or hobby brushed motors.

At this point, I think I have covered real EMCs sufficiently so that it would be wise to return to the original OO discussion and see how all the previous material fits into an OO hierarchy.

EMC Class Implementations

The first file that needs to be created in this example is the one that holds all the class definitions. I named this file emc.py to reflect both the project intent and the parent, or base class name. The file is shown below without any comments because I discuss the contents after the listing:

```
class EMC(object):
    def __init__(self, tau, period):
        self.tau = tau
        self.period = period
```

```
    def getTau(self):
        return self.tau

    def getPeriod(self):
        return self.period

    def __str__(self):
        return "tau = %s, period = %s" % (self.tau, self.period)
class Servo(EMC):
    def moveForward(self, n):
        print("Rotating the servo CW " + str(n) + " degrees")

    def moveBack(self, n):
        print("Rotating the servo CCW " + str(n) + " degrees")

class CRServo(EMC):
    def moveForward(self, n):
        print("Rotating the CR servo CW for " + str(n) + " seconds")

    def moveBack(self, n):
        print("Rotating the CR servo CCW for " + str(n) + " seconds")

class AnalogMotor(EMC):
    def moveForward(self, n):
        print("Rotating the analog motor CW for " + str(n) +
        " seconds")

    def moveBack(self, n):
        print("Rotating the analog motor CCW for " + str(n) +
        " seconds")
```

The file begins with the statement `class EMC(object):` where the reserved word indicates that a class definition follows. I used EMC for the base class name, as it again reflects the object being modeled. Note that I used all capitals for the name just because it was very short. Please note that all class names should begin with a capital letter. Python does not require this; it is strictly a convention, or style, which is followed by most developers, no matter which language they are using. The word "`object`" in the parentheses refers to the highest-level class in Python and is required for the base class definition.

This class has two attributes or descriptors, `tau` and `period`. I chose them because they are common to all the sub, or child, classes and are required operational parameters for the EMCs. They first appear in the special function, or method, named `__init__`. The word init is immediately preceded and followed by two underscores. The `init` method is called each time an object of the EMC class is created, or instantiated. It is also referred to, in OO terms, as the class constructor. The word "`self`" in the `init` method is a reference to itself. Therefore, when an expression, such as `self.tau = tau` is executed, the Python interpreter places a copy of the value passed by the `tau` argument into the object's `tau` parameter. It may seem a bit confusing, but remember, many different objects of the same class can be instantiated, each with its own value of `tau`, along with any other defined attributes

There are three methods in the EMC class definition. Two of them, `getTau` and `getPeriod`, are known as *getters* due to both their names and the function they serve: returning attribute values. There is a complementary method commonly know as a *setter* in which an attribute's value can be explicitly set. I chose not to include any setters in the class definition because I used the `init` method/ constructor to initialize them. Getters and setters are also collectively known as *mutators* because they allow attributes to be exposed outside of the class. Mutators are a key way a class maintains its encapsulation, meaning that internal attribute values can be set or retrieved only through these methods. Attributes are never exposed as global variables, which would mean any outside entity could change them. That would nullify a major OO class feature.

The other method in the class is `__str__`, which returns a string when called. This is a very handy feature when you wish to print or display an object. Ordinarily, Python would have no way of determining how to print a specific object other than its default method, which is to print the object's name and a hexadecimal number showing its physical memory location. That really is not very helpful when trying to examine an object. Incidentally, the default object's print behavior comes from the object class specified in the initial EMC class definition. The `__str__` method overrides the default behavior to create a usable object print display.

The three child or subclasses all inherit from the EMC class. This is done by including the EMC name as an argument at the start of the child class definition. For example, `class Servo(EMC)` sets the inheritance relationship between the Servo and the EMC classes. Technically, a Servo object "is an" EMC object. This "is a(n)" relationship is very important in creating an OO hierarchy. I do need to mention the other important OO relationship of "has a(n)," which refers to composition. That's where one object contains a reference to another object. Think of two classes, one modeling a car and the other an engine. Many of my beginning OO students make the novice mistake of making the Engine a sub-class of the Car class. This leads to the absurd statement that an engine is a car. That doesn't make sense; thus it is not inheritance. However, if you have the

Car class that contains a reference to the Engine class, then you can make the statement that a car has an engine, which makes perfect sense. Just keep in mind the "is a" and "has a" relationships, and if used in a sentence that makes sense, you have properly identified the correct relationship.

Each of the child classes has two methods, moveForward and moveBack. However, the methods have different strings they return, dependent upon the class they belong to. I used strings in lieu of trying to program actual EMC devices because it greatly simplified the code and still provided the class relationship example I desired. In a real-world example, there would be very specific code present using the tau and period attributes to control the servos and/or motors. Also notice that all the methods in the child class definitions have an integer argument n. This argument could represent either the number of degrees to rotate for the Servo class or the number of seconds to operate for the CRServo and AnalogMotor classes. The following test class file named EMCTest.py shows this very clearly. This time I added comments, since the program is fairly concise and relatively easy to understand.

```python
# need all the class definitions found in the emc.py file
from emc import EMC, Servo, CRServo, AnalogMotor

# instantiate a Servo object with tau = 0.5 and period = 1500 µsec
servo1 = Servo(0.5, 1500)

# instantiate a CRServo object with tau = 0.4 and period = 2000 µsec
crservo1 = CRServo1(0.4, 2000)

# instantiate an AnalogMotor object with tau = 1.0 and
  period = 2500 µsec
analog1 = AnalogMotor(1.0, 2500)

# call the Servo moveForward method using an n = 60 degrees
servo1.moveForward(60)
# get the tau for the servo1 object.  It was set to 0.4 in
  the constructor
t1 = servo1.getTau()
# display the tau value.  Need to convert the float value to a string
  for concatenation (+)
print("Servo tau = " + str(t1))

# call the CRServo moveBack method using an n = 2 seconds
crservo1.moveBack(2)
# print the crservo1 object.  Uses the __str__ method of the EMC class
print crservo1

# call the AnalogMotor moveForward method using an n = 3 seconds
analog1.moveForward(3)
```

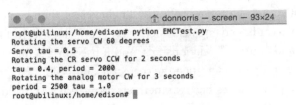

Figure 7-12 *EMCTest program result.*

```
# print both period and tau using the getters from the EMC class
# note that both getter arguments use the object name, analog1
print "period = %s tau = %s" % (EMC.getPeriod(analog1), EMC.
getTau(analog1))
```

Figure 7-12 is the result of running the EMCTest program, which is done by entering:

```
python EMCTest.py
```

You must have the emc.py file in the same directory as the EMCTest.py program, or Python will not be able to locate and load the class definitions.

This last program concludes this OO section. As you might imagine, I have only scratched the surface regarding OO programming. I would highly encourage you to take a formal course in any of the OO languages to gain valuable knowledge and experience.

Summary

This chapter provided an introduction to key OO concepts and to the Python programming techniques used to implement those concepts. I presented a relevant example using electronic motor controllers (EMC) to demonstrate how to create a Python inheritance hierarchy that could be used with the Edison controlling a dev board.

8

Hardware Interfaces

There are eight hardware interfaces available on the Edison according to Intel's product brochure. Some of these interfaces, such as GPIO and USB, have already been discussed in earlier chapters. This chapter will cover the remaining ones with some Python examples provided for two of the interfaces that you will likely use for projects. All the hardware interfaces with associated controllers are listed in Table 8-1.

Serial Protocols

Five of the eight hardware interfaces are bit serial and are used for communications and peripheral control. I already have used the USB 2.0 interface in earlier chapters to establish communications between the dev board and a laptop running a terminal program. Most readers will not attempt to create their own project USB interface because it is a fairly complex endeavor. There are many open-source and commercial applications available that will work quite well and, as a result, eliminate the need to create your own custom version.

All four serial protocols described in the following sections are implemented in hardware, meaning there is actual silicon dedicated to performing the protocol's functions. This is the most efficient and fastest way to provide these serial interfaces, but it is not the only way. These same serial interfaces can be implemented using uncommitted GPIO pins and software. This would provide nearly the same functionality but would not be as fast as compared to the hardware implementation. The term "bit-banging" is often used to describe this approach. Sometimes you must use bit-banging when the hardware is not available.

Name	# Controllers	Type	Remarks
USB 2.0	1	Serial	With OTG
UART	2	Serial	Communications. One controller with full RX/TX duplex control
I2C	2	Serial	Peripheral control
SPI	1	Serial	Peripheral control
I2S	1	Serial	Digital Audio
GPIO	40	Parallel	Digital I/O lines
SD Card	1	Parallel	Large memory block control
Clock	1	Parallel	32 KHz and 19.2 MHz internal clocks

Table 8-1   *Edison Hardware Interfaces*

UART Serial Protocol

UART is short for *Universal Asynchronous Receive Transmit* and is a bit serial protocol that uses pins 0 and 1 on the dev board. Figure 8-1 shows a diagram of the protocol. The block named Node in the figure represents any compatible device with a UART interface.

This protocol needs no clock signal, as is indicated by the adjective "asynchronous" in the name. The Edison transmits data on pin 1 named TX and receives data on pin 0, named RX. There is also no concept of a master or slave in this protocol, since it is used primarily for data communications instead of control, which is the focus of both the SPI and I2C interfaces.

I used a Python library named pyserial to demonstrate a simple UART protocol using the same serial interface that connects to the USB communications

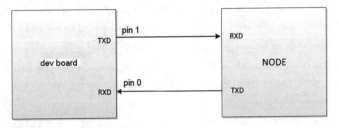

Figure 8-1   *UART block diagram.*

port, which is /dev/ttyMFD2. The program just displays a series of numbers using an object from the pyserial class to process the data. You will need to perform the following steps in order to load and install the pyserial library:

1. Enter the following command to install an application named easy_install. Remember to be at the root level.

    ```
    wget https://bootstrap.pypa.io/ez_setup.py - -no-
    check-certificate -O - | python
    ```

2. Next enter the following to load the pyserial library:

    ```
    easy_install -U pyserial
    ```

Once the library is loaded, use nano to create the following program named uart.py.

```
#!/usr/bin/python
# uart.py
# D. J. Norris 2/2015
# The code is in the public domain
import time
import serial
serialPort = '/dev/ttyMFD2'

# this is the serial UART object that is set to 115200 baud
ser = serial.Serial(serialPort, 115200)
count = 0

# loop to display a series of integers
for x in range(0,14):
    ser.write(str(count))
    ser.write('/n') # new line
    ser.write('/r') # carriage return
    count += 1 # increment the count
print "All done"
```

Figure 8-2 shows the result of running this simple program, which you do by entering:

```
python uart.py
```

You could also use pins 0 and 1 on the dev board for another UART connection to a computer running a separate terminal program. The logical device name associated with these pins is /dev/ttyMFD1. Just ensure that pin 0 is connected to the external computer's transmit or TX lead. Pin 1, likewise, must be connected to the external computer's receive or RX line.

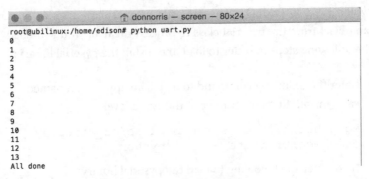

Figure 8-2   *uart.py program results.*

NOTE   *The signal levels from the dev board are 5 V and 0 V, which may not be compatible with the external computer's UART. Connecting the dev board to the external computer will not harm either one, but there just won't be a communications link established. My advice is to try it and see if it works. I would recommend using a single board computer, such as the Rasberry Pi, as the external computer. I know its UART pins will function properly with the dev board*

I2C Serial Protocol

The next serial protocol that I will discuss is the Inter-Integrated Circuit interface or I2C (pronounced "eye-two-cee" or "eye-squared-cee"), which is a synchronous serial data link. Figure 8-3 is a block diagram of the I2C interface showing one master and one slave. This configuration is also known as a *multidrop* or *bus* network.

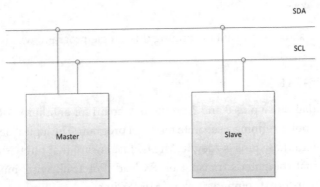

Figure 8-3   *I2C block diagram.*

Signal Name	Description	Dev Board Name
SCL	Clock	SCL
SDA	Data	SDA

Table 8-2   *I2C Signal Lines*

I2C supports more than one master as well as multiple slaves. This protocol was created by Philips Semiconductors (now NXP Semiconductors) in 1982 and is a very mature technology, meaning it is extremely reliable. Only two lines are used: SCLK for serial clock and SDA for serial data. Table 8-2 shows the dev board names for both the clock and data lines.

I2C Software

I strongly recommend that you perform the next few steps before attempting to duplicate the I2C hardware example. Again, you must be at the root level for these steps.

```
apt-get update
apt-get upgrade    (Be patient, this could take a while)
```

These next commands ensure that the Wheezy Debian distribution is current with useful I2C tools and utilities.

```
apt-get install python-smbus
apt-get install i2c-tools
```

These commands load some applications that allow you to check the address of attached I2C modules among other things. The i2c-tools package contains the i2cdetect application that will display all the I2C devices connected on the bus.

The python-smbus library package contains additional software to access I2C devices from Python. Note that the I2C hardware example in this section uses mraa to access Python, but I do include a brief example of how to use the python-smbus library for your information.

I2C and the BMP085 Sensor

I decided to use a BMP085 combined pressure and temperature sensor manufactured by the Bosch Corp. The BMP085 uses the I2C bus to send and receive both commands and data between itself and a microprocessor, which, in this case, will be the Edison mounted on a dev board. Figure 8-4 is a block diagram excerpted from the BMP085 datasheet, illustrating how the interconnections should be made between the sensor and microprocessor.

Please note that you should not use the two resistors, shown in the diagram, that connect the SCL and SDA data lines to the V_{CC} supply. The Edison already has these pull-up resistors installed and adding the external ones will cause the I2C signals to fail.

Figure 8-5 is the physical setup showing the BMP085 mounted on a solderless breadboard with four jumpers between it and the dev board. The setup is quick and relatively simple to accomplish. It becomes more interesting when I discuss the software needed to operate the sensor in the next section

BMP085 Test Software

I used a class named BMP085 both to model the sensor and to reinforce the class concepts I introduced in Chapter 7. It turns out the BMP085 is a relatively

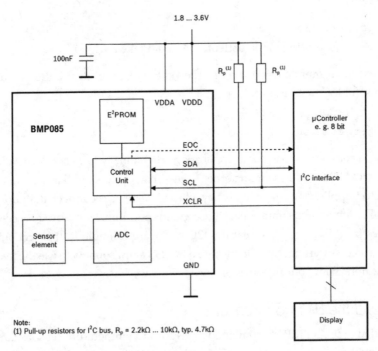

Figure 8-4 *BMP085 interconnection block diagram.*

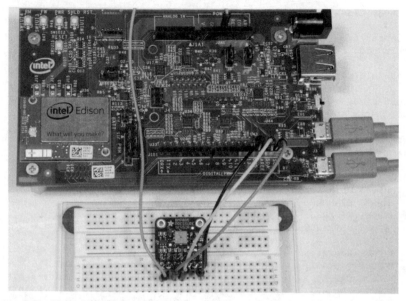

Figure 8-5 *Physical BMP085 test setup.*

sophisticated sensor with an onboard memory containing calibration constants, which permit the calculation of precise temperature and pressure measurements from raw sensor readings. The class definition contains all the attributes and methods needed to perform these calculations. I should note that I chose to perform only the temperature calculations because that was sufficient to demonstrate how to properly use this I2C sensor. The pressure and temperature calculations are quite similar, and I do not believe you should have a difficult time implementing the pressure calculations, if you so chose.

The following BMP085 class listing is quite long because there are a lot of calibration constants to be handled as well as the associated calculations. Also, the listing contains extensive debug statements and conditions spread throughout the class, which will help you understand how the class works. I adapted this class from one created by the bright folks at Adafruit Industries. Their version is based on their I2C class library, while this one is designed to work with the mraa library. The Adafruit BMP085 class version will not work with the mraa library without some moderate changes. I would also recommend you read the BMP085 datasheet if you wish to learn how the calibration procedure functions.

```python
# BMP085 Class for mrra
# D. J. Norris 2/2015
# Adapted from the Adafruit BMP085 class definition
# This code is in the public domain

#!/usr/bin/python

import time
import mraa as m

class BMP085(object):
  i2c = None

  # Operating Modes
  __BMP085_ULTRALOWPOWER       =       0
  __BMP085_STANDARD            =       1
  __BMP085_HIGHRES             =       2
  __BMP085_ULTRAHIGHRES        =       3

  # BMP085 Registers
  __BMP085_CAL_AC1             = 0xAA  # R   Calibration data (16 bits)
  __BMP085_CAL_AC2             = 0xAC  # R   Calibration data (16 bits)
  __BMP085_CAL_AC3             = 0xAE  # R   Calibration data (16 bits)
  __BMP085_CAL_AC4             = 0xB0  # R   Calibration data (16 bits)
  __BMP085_CAL_AC5             = 0xB2  # R   Calibration data (16 bits)
  __BMP085_CAL_AC6             = 0xB4  # R   Calibration data (16 bits)
  __BMP085_CAL_B1              = 0xB6  # R   Calibration data (16 bits)
  __BMP085_CAL_B2              = 0xB8  # R   Calibration data (16 bits)
  __BMP085_CAL_MB              = 0xBA  # R   Calibration data (16 bits)
  __BMP085_CAL_MC              = 0xBC  # R   Calibration data (16 bits)
  __BMP085_CAL_MD              = 0xBE  # R   Calibration data (16 bits)
  __BMP085_CONTROL             = 0xF4
  __BMP085_TEMPDATA            = 0xF6
  __BMP085_PRESSUREDATA        = 0xF6
  __BMP085_READTEMPCMD         = 0x2E
  __BMP085_READPRESSURECMD     = 0x34

  # Private Fields
  _cal_AC1 = 0
  _cal_AC2 = 0
  _cal_AC3 = 0
  _cal_AC4 = 0
  _cal_AC5 = 0
  _cal_AC6 = 0
  _cal_B1 = 0
  _cal_B2 = 0
  _cal_MB = 0
  _cal_MC = 0
  _cal_MD = 0

  # Constructor
  def __init__(self, address=0x77, bus = 0, mode=1, debug=False):
    self.i2c = m.I2c(bus)
    self.i2c.address(address)
```

```
      self.bus = bus
      self.address = address
      self.debug = debug
      # Make sure the specified mode is in the appropriate range
      if ((mode < 0) | (mode > 3)):
        if (self.debug):
          print "Invalid Mode: Using STANDARD by default"
          self.mode = self.__BMP085_STANDARD
      else:
        self.mode = mode
      # Read the calibration data
      self.readCalibrationData()

  def readS16(self, register):
    "Reads a signed 16-bit value"
    hi = self.i2c.readReg(register)
    shi = (hi + 2**7) % 2**7 - 2**7
    lo = self.i2c.readReg(register+1)
    i = (shi << 8) + lo
    return i

  def readU16(self, register):
    "Reads an unsigned 16-bit value"
    hi = self.i2c.readReg(register)
    lo = self.i2c.readReg(register+1)
    i = (hi << 8) + lo
    return i

  def readCalibrationData(self):
    "Reads the calibration data from the IC"
    self._cal_AC1 = self.readS16(self.__BMP085_CAL_AC1)   # INT16
    self._cal_AC2 = self.readS16(self.__BMP085_CAL_AC2)   # INT16
    self._cal_AC3 = self.readS16(self.__BMP085_CAL_AC3)   # INT16
    self._cal_AC4 = self.readU16(self.__BMP085_CAL_AC4)   # UINT16
    self._cal_AC5 = self.readU16(self.__BMP085_CAL_AC5)   # UINT16
    self._cal_AC6 = self.readU16(self.__BMP085_CAL_AC6)   # UINT16
    self._cal_B1  = self.readS16(self.__BMP085_CAL_B1)    # INT16
    self._cal_B2  = self.readS16(self.__BMP085_CAL_B2)    # INT16
    self._cal_MB  = self.readS16(self.__BMP085_CAL_MB)    # INT16
    self._cal_MC  = self.readS16(self.__BMP085_CAL_MC)    # INT16
    self._cal_MD  = self.readS16(self.__BMP085_CAL_MD)    # INT16

    if (self.debug):
      self.showCalibrationData()

  def showCalibrationData(self):
    "Displays the calibration values for debugging purposes"
    print "DBG: AC1 = %6d" % (self._cal_AC1)
    print "DBG: AC2 = %6d" % (self._cal_AC2)
    print "DBG: AC3 = %6d" % (self._cal_AC3)
    print "DBG: AC4 = %6d" % (self._cal_AC4)
    print "DBG: AC5 = %6d" % (self._cal_AC5)
    print "DBG: AC6 = %6d" % (self._cal_AC6)
    print "DBG: B1  = %6d" % (self._cal_B1)
```

```
      print "DBG: B2   = %6d" % (self._cal_B2)
      print "DBG: MB   = %6d" % (self._cal_MB)
      print "DBG: MC   = %6d" % (self._cal_MC)
      print "DBG: MD   = %6d" % (self._cal_MD)

def readRawTemp(self):
    "Reads the raw (uncompensated) temperature from the sensor"
    self.i2c.writeReg(self.__BMP085_CONTROL,
self.__BMP085_READTEMPCMD)
    time.sleep(1.0)  # Wait one second
    raw = self.i2c.readWordReg(self.__BMP085_TEMPDATA)
    print "Raw temp = " + str(raw)

    if (self.debug):
      print "DBG: Raw Temp: 0x%04X (%d)" % (raw & 0xFFFF, raw)

    return raw

def readRawPressure(self):
    "Reads the raw (uncompensated) pressure level from the sensor"
    self.i2c.write8(self.__BMP085_CONTROL,   self.__BMP085_
READPRESSURECMD+(self.mode << 6))
    if (self.mode == self.__BMP085_ULTRALOWPOWER):
      time.sleep(0.005)
    elif (self.mode == self.__BMP085_HIGHRES):
      time.sleep(0.014)
    elif (self.mode == self.__BMP085_ULTRAHIGHRES):
      time.sleep(0.026)
    else:
      time.sleep(0.008)
    msb = self.i2c.readU8(self.__BMP085_PRESSUREDATA)
    lsb = self.i2c.readU8(self.__BMP085_PRESSUREDATA+1)
    xlsb = self.i2c.readU8(self.__BMP085_PRESSUREDATA+2)
    raw = ((msb << 16) + (lsb << 8) + xlsb) >> (8 - self.mode)

    if (self.debug):
      print "DBG: Raw Pressure: 0x%04X (%d)" % (raw & 0xFFFF, raw)

    return raw

  def readTemperature(self):
    "Gets the compensated temperature in degrees Celcius"
    UT = 0
    X1 = 0
    X2 = 0
    B5 = 0
    temp = 0.0
    # Read raw temp before aligning it with the calibration values
    UT = self.readRawTemp()
    X1 = ((UT - self._cal_AC6) * self._cal_AC5) >> 15
    X2 = (self._cal_MC << 11) / (X1 + self._cal_MD)
    B5 = X1 + X2
    temp = ((B5 + 8) >> 4) / 10.0
```

```
    if (self.debug):
      print "DBG: Calibrated temperature = %f C" % temp

    return temp

def readPressure(self):
  "Gets the compensated pressure in Pascal"
  UT = 0
  UP = 0
  B3 = 0
  B5 = 0
  B6 = 0
  X1 = 0
  X2 = 0
  X3 = 0
  p = 0
  B4 = 0
  B7 = 0
  UT = self.readRawTemp()
  UP = self.readRawPressure()

  # You can use the datasheet values to test the conversion results
  # dsValues = True
  dsValues = False
  if (dsValues):
    UT = 27898
    UP = 23843
    self._cal_AC6 = 23153
    self._cal_AC5 = 32757
    self._cal_MB = -32768;
    self._cal_MC = -8711
    self._cal_MD = 2868
    self._cal_B1 = 6190
    self._cal_B2 = 4
    self._cal_AC3 = -14383
    self._cal_AC2 = -72
    self._cal_AC1 = 408
    self._cal_AC4 = 32741
    self.mode = self.__BMP085_ULTRALOWPOWER

    if (self.debug):
      self.showCalibrationData()

  # True Temperature Calculations
  X1 = ((UT - self._cal_AC6) * self._cal_AC5) >> 15
  X2 = (self._cal_MC << 11) / (X1 + self._cal_MD)
  B5 = X1 + X2

  if (self.debug):
    print "DBG: X1 = %d" % (X1)
    print "DBG: X2 = %d" % (X2)
    print "DBG: B5 = %d" % (B5)
    print "DBG: True Temperature = %.2f C" % (((B5 + 8) >> 4) / 10.0)
```

```
# Pressure Calculations
B6 = B5 - 4000
X1 = (self._cal_B2 * (B6 * B6) >> 12) >> 11
X2 = (self._cal_AC2 * B6) >> 11
X3 = X1 + X2
B3 = (((self._cal_AC1 * 4 + X3) << self.mode) + 2) / 4

if (self.debug):
  print "DBG: B6 = %d" % (B6)
  print "DBG: X1 = %d" % (X1)
  print "DBG: X2 = %d" % (X2)
  print "DBG: X3 = %d" % (X3)
  print "DBG: B3 = %d" % (B3)

X1 = (self._cal_AC3 * B6) >> 13
X2 = (self._cal_B1 * ((B6 * B6) >> 12)) >> 16
X3 = ((X1 + X2) + 2) >> 2
B4 = (self._cal_AC4 * (X3 + 32768)) >> 15
B7 = (UP - B3) * (50000 >> self.mode)

if (self.debug):
  print "DBG: X1 = %d" % (X1)
  print "DBG: X2 = %d" % (X2)
  print "DBG: X3 = %d" % (X3)
  print "DBG: B4 = %d" % (B4)
  print "DBG: B7 = %d" % (B7)

if (B7 < 0x80000000):
  p = (B7 * 2) / B4
else:
  p = (B7 / B4) * 2

if (self.debug):
  print "DBG: X1 = %d" % (X1)

X1 = (p >> 8) * (p >> 8)
X1 = (X1 * 3038) >> 16
X2 = (-7357 * p) >> 16

if (self.debug):
  print "DBG: p  = %d" % (p)
  print "DBG: X1 = %d" % (X1)
  print "DBG: X2 = %d" % (X2)

p = p + ((X1 + X2 + 3791) >> 4)

if (self.debug):
  print "DBG: Pressure = %d Pa" % (p)

return p

def readAltitude(self, seaLevelPressure=101325):
  "Calculates the altitude in meters"
```

```
    altitude = 0.0
    pressure = float(self.readPressure())
    altitude = 44330.0 * (1.0 - pow(pressure / seaLevelPressure,
0.1903))

    if (self.debug):
      print "DBG: Altitude = %d" % (altitude)

    return altitude

    return
```

The BMP085 class must be instantiated in a test class in order to use the sensor for temperature and pressure measurements. That is the purpose of the following test class named Test_BMP085.py, which is a slightly modified version of the one included with the Adafruit BMP085 software library.

```
#!/usr/bin/python

import mraa as m
from BMP085 import BMP085

# Test_BMP085.py
# Example Code
#

# Initialize the BMP085 and use STANDARD mode (default value)
# bmp = BMP085(0x77, debug=True)
bmp = BMP085(0x77) # the argument is not needed as it is hard-coded in
                   the ctor

# To specify a different operating mode, uncomment one of the follow-
ing:
# bmp = BMP085(0x77, 0)   # ULTRALOWPOWER Mode
# bmp = BMP085(0x77, 1)   # STANDARD Mode
# bmp = BMP085(0x77, 2)   # HIRES Mode
# bmp = BMP085(0x77, 3)   # ULTRAHIRES Mode

temp = bmp.readTemperature()

# Read the current barometric pressure level
#pressure = bmp.readPressure()

# To calculate altitude based on an estimated mean sea level pressure
# (1013.25 hPa) call the function as follows, but this won't be very
  accurate
# altitude = bmp.readAltitude()

# To specify a more accurate altitude, enter the correct mean sea
  level pressure level.
# For example, if the current pressure level is 1023.50 hPa
# enter 102350 since we include two decimal places in the integer value
```

```
# altitude = bmp.readAltitude(102350)

print "Temperature:    %.2f C" % temp
# print "Pressure:     %.2f hPa" % (pressure / 100.0)
# print "Altitude:     %.2f" % altitude
```

The above class is run by entering the following at `root` level:

```
python Test_BMP085.py
```

Figure 8-6 shows the result of this command. The measured temperature of 22.2° C is very close to 72° F, which was the ambient room temperature during this test.

Brief Tests Using the SMBus Utilities

SMBus is an abbreviation for *System Management Bus* and is essentially a clone of the I2C bus. It was introduced by Intel in 1995 as an improvement to the I2C bus; however, they are almost identical in any practical sense. SMB uses the same data and control lines as does I2C and applies the same names to those lines.

There are several SMBus Linux libraries available with the python-smbus package, which you should already have downloaded and installed if you followed my steps detailed earlier in this chapter.

The library contains a useful utility named i2cdetect, which can aid you in setting up an I2C device. Enter the following command to examine the functionalities that your connected I2C device will support:

```
i2cdetect -F 1
```

Figure 8-7 shows the results of this command for the connected BMP085 sensor. As you can see, the BMP085 supports most of the SMBus commands except for a few, including the SMBus Quick Command. Unfortunately, the Quick Command is used to request that the device report its own address. I had

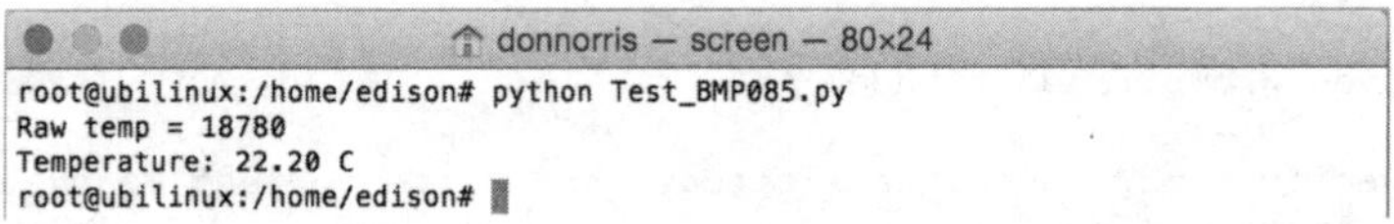

```
● ● ●              ⌂ donnorris — screen — 80×24
root@ubilinux:/home/edison# python Test_BMP085.py
Raw temp = 18780
Temperature: 22.20 C
root@ubilinux:/home/edison# ▮
```

Figure 8-6   *Test_BMP085 program results.*

```
root@ubilinux:/home/edison# i2cdetect -F 1
Functionalities implemented by /dev/i2c-1:
I2C                              yes
SMBus Quick Command              no
SMBus Send Byte                  yes
SMBus Receive Byte               yes
SMBus Write Byte                 yes
SMBus Read Byte                  yes
SMBus Write Word                 yes
SMBus Read Word                  yes
SMBus Process Call               no
SMBus Block Write                no
SMBus Block Read                 no
SMBus Block Process Call         no
SMBus PEC                        no
I2C Block Write                  yes
I2C Block Read                   yes
root@ubilinux:/home/edison# █
```

Figure 8-7   *i2cdetect utility program results.*

to modify Figure 8-8 slightly to show the I2C address as if the BMP085 had correctly responded to the Quick Command for display address.

```
i2cdetect -y 1
```

The hex address 0x77 shown in the figure is the I2C address programmed into the BMP085 sensor by the manufacturer.

SPI Serial Protocol

SPI is short for the *Serial Peripheral Interface*, which is shown in the block diagram in Figure 8-9. The SPI interface (pronounced "spy" or "ess-pee-eye") is a synchronous serial data link. A clock signal is needed because it is synchronous. It is also a full duplex protocol, meaning data can be simultaneously sent and received between the host and slave. SPI is also referred to as a *Synchronous Serial Interface* (SSI), or a *4-wire serial bus*.

```
root@ubilinux:/home/edison# i2cdetect -y -r 1
     0  1  2  3  4  5  6  7  8  9  a  b  c  d  e  f
00:          -- -- -- -- -- -- -- -- -- -- -- --
10: -- -- -- -- -- -- -- -- -- -- -- -- -- -- -- --
20: UU UU UU UU -- -- -- -- -- -- -- -- -- -- -- --
30: -- -- -- -- -- -- -- -- -- -- -- -- -- -- -- --
40: -- -- -- -- -- -- -- -- -- -- -- -- -- -- -- --
50: -- -- -- -- -- -- -- -- -- -- -- -- -- -- -- --
60: -- -- -- -- -- -- -- -- -- -- -- -- -- -- -- --
70: -- -- -- -- -- -- -- 77
root@ubilinux:/home/edison# █
```

Figure 8-8   *i2cdetect address command program results.*

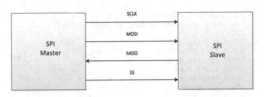

Figure 8-9 *SPI block diagram.*

The four interconnecting signal lines between the SPI host and SPI slave shown in Figure 8-9 are explained in Table 8-3.

Loopback Test Program

A loopback test is probably the simplest Python program that can be created to demonstrate the SPI interface. A loopback is set up by connecting the SPI output pin 11 (MOSI) directly to the SPI input pin 12 (MISO). A simple jumper wire is sufficient to make this connection on the dev board's header.

I created a fairly straightforward program that sends 17 integers to the SPI output. The SPI input receives those integers because of the loopback, and they are displayed on a terminal screen, one at a time, with 0.5 seconds between integers.

In order to make the following program run, I had to switch my Edison module mounted on the dev board from one with a Debian distribution to a module flashed with the latest Poky distribution. I had mentioned in an earlier chapter that the latest versions of Poky contain the most updated mraa library version. I am not sure why SPI doesn't work well with the mraa Debian version, but it seems to be OK with the current Poky version. In any case, the following program named TestSPI.py worked without any issues using Poky's mraa.

```
# TestSPI.py
# D. J. Norris 2/2015
# Code is in the public domain
# Ensure Arduino board pins 11 & 12 are connected together

#!/usr/bin/python
```

Signal Name	Description	Dev Board Pin #
SCLK	Clock	13
MOSI	Master Out Slave In	11
MISO	Master In Slave Out	12
SS	Slave Select	4

Table 8-3 *SPI Signal Lines*

```
import time
import mraa as m
# Bus 5 from /dev/spi5.1
spi = m.Spi(5)
count = 0
while True:
    try:
        num = len(str(count)) # get string length for display
        recNum = spi.write(str(count)) # write returns argument
        print recNum[:num] # simple loopback
        time.sleep(0.5) # wait a short time
        count += 1
        if count > 16:
            break # breakout of the loop after count = 16
    except(KeyboardInterrupt, SystemExit):
        raise
print "All done"
```

Note that when the SPI `write` method is called, it automatically returns the value it wrote. That is why there is no `read` method required for the SPI protocol.

The loopback program is called by entering:

```
python TestSPI.py
```

Figure 8-10 shows the program results, which are only a series of integers reflecting what is both sent and received via the SPI interface.

I will next demonstrate a more interesting project in which I interface the dev board to a real SPI-controlled device.

Figure 8-10   *TestSPI program results.*

Connecting an MCP3008 to the Dev Board

I will be using a Microchip model MCP3008 to demonstrate an SPI hardware interface with the dev board. The MCP3008 is a 4-channel, 10-bit analog-to-digital converter (ADC) with a built-in SPI serial interface. There are a minimum of four data lines used with the SPI interface, which are listed in Table 8-4 along with the names associated with the master (dev board) and the slave (MCP3008) devices.

Figure 8-11 is a simplified interconnection block diagram showing the principal components used in this SPI data link. There are usually two shift registers involved in the data link, as shown in the figure. These registers may be hardware or software, depending upon the devices involved. The Edison implements its shift register in software, while the MCP3008 has a hardware shift register. In either case, the two shift registers form what is known as an interchip circular buffer arrangement that is the heart of the SPI.

Data communications is initiated by the master, which begins by selecting the required slave. The Edison selects the MCP3008 by bringing the SS line to a low state or 0 V DC. During each clock cycle, the master sends a bit to the slave, which reads it from the MOSI line. Concurrently, the slave sends a bit to the master, which reads it from the MISO line. This operation is known as full duplex communication, i.e., simultaneous reading and writing between master and slave.

The clock frequency used is dependent primarily upon the slave's response speed. The MCP3008 can easily handle bit rates up to 3.6 MHz, if powered at 5 V. In this test, the MCP3008 will be powered by the dev board's 3.3-V power supply, which probably sets the bit rate at about 2.5 MHz for this setup. That lower rate is still quite sufficient for this test.

The first clock pulse received by the MCP3008, with its CS held low and Din high, constitutes the start bit. The SGL/DIFF bit follows next and then three bits that represent the selected channel(s). After these five bits have been received, the MCP3008 will sample the analog voltage during the next clock cycle.

Master Device—Dev Board	Slave Device—MCP3008	Remarks
SCLK - pin 13	CLK - pin 13	Clock
MOSI - pin 11	Din - pin 11	Master Out Slave In
MISO - pin 12	Dout - pin 12	Master In Slave Out
SS - pin 10	CS - pin 10	Slave Select

Table 8-4 *Dev Board to MCP3008 SPI Data Line Descriptions*

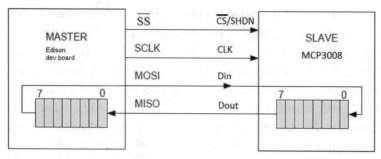

Figure 8-11 *SPI simplified block diagram.*

The MCP3008 then outputs what is known as a low null bit that is disregarded by the Edison. The following 10 bits, each sent on a clock cycle, are the ADC value with the most significant bit (MSB) sent first down to the least significant bit (LSB) sent last. The Edison will then put the MCP3008 CS pin high, ending the ADC process

Testing the MCP3008 with the Edison

First ensure that the MCP3008 is connected to the dev board, as shown in the Figure 8-12 interconnection diagram. There is a temporary test setup on the left

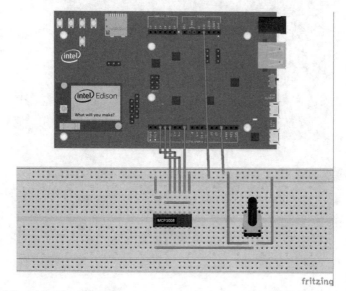

Figure 8-12 *Interconnection diagram.*

side of the breadboard consisting of a potentiometer connected between 3.3 V DC and ground. The ADC channel 0 is connected to the tap allowing a variable voltage that can be used in the test. The physical setup is shown in Figure 8-13.

The test program is named TestSPI_ADC.py and will generate a continuous stream of ADC values from a source connected to channel 0. The code follows the ADC configuration and SPI protocols discussed earlier. The code, while extensively modified, is based upon the sample code available from the Learn .Adafruit.com website in their discussion of the MCP3008.

```
#!/usr/bin/python

# TestSPI_ADC.py
#  D. J. Norris 2/2015
# This code is in the public domain
# Adapted from Adafruit Industries source code

import time
import mraa as m

# turn on debug for testing
 DEBUG = 1
```

Figure 8-13 *MCP3008 and dev board test setup.*

```
# define a function to read the MCP3008 ADC value
def adc(chan, clock, mosi, miso, cs):
    if((chan < 0) or (chan > 7)):
        return -1
    cs.write(True)
    clock.write(False)
    cs.write(False)
    cmd = chan
    cmd |= 0x18
    cmd <<= 3

    for i in range(5):
        if(cmd & 0x80):
            mosi.write(True)
        else:
            mosi.write(False)
        cmd <<= 1
        clock.write(True)
        clock.write(False)
    result = 0
    for i in range(12):
        clock.write(True)
        clock.write(False)
        result <<= 1
        if(miso.read()):
            result |= 0x1
    cs.write(True)
    result >>= 1
    return result

#These pin definitions are set to work with the test circuit
SPICLK = m.Gpio(4)
SPIMISO = m.Gpio(2)
SPIMOSI = m.Gpio(3)
SPICS = m.Gpio(7)


# Set pins for direction
SPICLK.dir(m.DIR_OUT)
SPIMISO.dir(m.DIR_IN)
SPIMOSI.dir(m.DIR_OUT)
SPICS.dir(m.DIR_OUTPUT)

# use channel 0 as a test input
channel = 0

while True:
    #read an ADC value
    adc_value = adc(channel, SPICLK, SPIMOSI, SPIMISO, SPICS)

    if DEBUG:
        print "value = ", adc_value
    #wait a second and repeat
    time.sleep(1)
```

Figure 8-14 is screenshot of a portion of the program output with the analog voltage count showing an average of 510 for 23 samples. The actual input voltage to the ADC chip was 1.73 V, as measured with an uncalibrated volt/ohm meter (VOM). The predicted ADC count for this voltage would be 512 based on the following calculation.

Predicted count = (Actual voltage × 1023) / Full-scale voltage = 511.5
511.5 was rounded to 512 because ADC counts are only integers.

This is very close to the average measured value of 510. This is a difference of only 2 counts, or +0.2%, which is quite typical for this ADC type.

The sharp-eyed reader may have noticed that the program did not use the built-in SPI functionality but instead implemented a "bit-banged" interface that was mentioned earlier. This approach was taken because the Poky SPI class did not work with this chip, and I wanted to provide an actual SPI-type project. It really makes no functional difference which approach is taken as long as the chip works as expected.

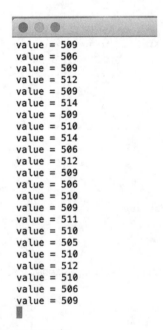

Figure 8-14 *TestSPI_ADC program results.*

I would also like to alert you that the above circuit and software will be used in the next chapter, in which I describe a temperature monitoring system. You might like to keep this circuit together if you intend to duplicate the next chapter's project.

I2S Serial Protocol

This is the last of the serial protocols to be discussed in this chapter. I2S is designed to handle only digital audio bit streams. I2S is short for *Integrated Interchip Sound* and was created by Philips Semiconductors in 1986 just five years after they introduced the I2C protocol. I2S is not related at all to I2C, as it was designed solely to process digitized audio samples and not for I2C control/communications purposes. It is not my intention to demonstrate any I2S example other than to mention that it does need to connect to specific chip sets in order to process digitized audio signals. The following are two chip sets that are available to work with the Edison in a digital audio project:

1. Wolfson Microelectronics WM8804

2. NXP UDA1380

The I2S protocol functions with pulse code modulation (PCM) audio data, which is sampled at 44.1 KHz, with 2 bytes or 16 bits representing the content of two stereo channels. That requirement results in an overall 1.411-MHz bit rate.

The I2S bus typically has three lines:

1. Word select clock

2. Bit clock

3. One multiplexed data line

The *word select clock* is used to specify to a receiving device whether the left or right channel stereo data is on the data line. The *bit clock* establishes synchronous communication between the host, which originates the audio data, and the receiving device. Finally, the *data line* sends the actual PCM data to the device.

According to the latest the *Edison Hardware Guide*, the I2S interface is available on pins 50, 52, 54, and 56. However, there are eight operational modes listed in the *Guide* that have not been verified to date. What I gather from this remark is that trying to develop a working I2S interface with the Edison might be problematic at this stage in its product development. I would recommend that the current Edison not be used to implement an I2S interface, at least until Intel assures us that the I2S modes work as planned and there is good supporting software available.

Parallel Protocols

Table 8-1 listed three parallel protocols, the GPIO, SD card, and clock. Of these three, you really will be using only the GPIO for your projects. I will briefly discuss each one in the next sections.

GPIO

GPIO refers to the 40 pins available on an integrated-curcuit board, most of which can function in two or more modes. The simplest mode is the one in which a pin can act either as a digital input or output, but not both at the same time. I have previously shown you various projects where GPIO pins either check for high and low voltage levels as inputs, or output highs or lows as needed. GPIO pins can be controlled in a variety of ways, including using the Arduino IDE or through the mraa library with an appropriate language, such as C/C++ or Python. I haven't discussed it yet, but pins may also be controlled directly from the Linux command line using export statements. These statements basically treat each pin as a read/write file with various attributes and behaviors. There is a good deal of information available on the Web regarding how to manipulate GPIO from the command line for those readers interested in pursuing this approach.

I believe using the mraa library along with the Python language provides a clean abstraction with which to handle most GPIO tasks. It is generally true that you can meet most programming requirements in a variety of ways, but I like to choose the simpler path, which gives me more time to handle other issues that always arise in a project.

One general proviso that is worth mentioning is that select pins are designated for special functions and should not be used for routine GPIO, if possible. On the Arduino dev board, pins 3, 5, 6, 9, 10, and 11 provide PWM functions as well as GPIO. I would avoid using those pins if there were a foreseeable need for PWM sometime in the future. Likewise, there are other functions some pins can do that others cannot. I would strongly suggest reviewing Table 8-5 to check pin function. This table also has a good cross-check for pin designations for both the Arduino dev board and the mini-breakout board.

Edison Pin (Linux)	Arduino Dev Board	Mini Breakout	mraa Number	Pinmode0	Pinmode1
GP12	3	J18-7	20	GPIO-12	PWM0
GP13	5	J18-1	14	GPIO-13	PWM1
GP14	A4	J19-9	36	GPIO-14	
GP15		J20-7	48	GPIO-15	
GP19		J18-6	19	GPIO-19	I2C-1-SCL
GP20		J17-8	7	GPIO-20	I2C-1-SDA
GP27		J17-7	6	GPIO-27	I2C-6-SCL
GP28		J17-9	8	GPIO-28	I2C-6-SDA
GP40	13	J19-10	37	GPIO-40	SSP2_CLK
GP41	10	J20-10	51	GPIO-41	SSP2_FS
GP42	12	J20-9	50	GPIO-42	SSP2_RXD
GP43	11	J19-11	38	GPIO-43	SSP2_TXD
GP44	A0	J19-4	31	GPIO-44	
GP45	A1	J20-4	45	GPIO-45	
GP46	A2	J19-5	32	GPIO-46	
GP47	A3	J20-5	46	GPIO-47	
GP48	7	J19-6	33	GPIO-48	
GP49	8	J20-6	47	GPIO-49	
GP77		J19-12	39	GPIO-77	SD
GP78		J20-11	52	GPIO-78	SD
GP79		J20-12	53	GPIO-79	SD
GP80		J20-13	54	GPIO-80	SD
GP81		J20-14	55	GPIO-81	SD
GP82		J19-13	40	GPIO-82	SD
GP83		J19-14	41	GPIO-83	SD
GP84		J20-8	49	GPIO-84	SD
GP109		J17-11	10	GPIO-109	SPI-5-SCK
GP110		J18-10	23	GPIO-110	SPI-5-CS0
GP111		J17-10	9	GPIO-111	SPI-5-CS1
GP114		J18-11	24	GPIO-114	SPI-5-MISO
GP115		J17-12	11	GPIO-115	SPI-5-MOSI
GP128	2	J17-14	13	GPIO-128	UART-1-CTS

Table 8-5 *Edison Pin Descriptions*

Edison Pin (Linux)	Arduino Dev Board	Mini Breakout	mraa Number	Pinmode0	Pinmode1
GP129	4	J18-12	25	GPIO-129	UART-1-RTS
GP130	0	J18-13	26	GPIO-130	UART-1-RX
GP131	1	J19-8	35	GPIO-131	UART-1-TX
GP134		J20-3	44		
GP135		J17-5	4	GPIO-135	UART
GP165	A5	J18-2	15	GPIO-165	
GP182	6	J17-1	0	GPIO-182	PWM2
GP183	9	J18-8	21	GPIO-183	PWM3

Table 8-5 *Edison Pin Descriptions* (Continued)

SD Card Interface

The Edison has eight parallel lines committed to the SD card interface. An SD interface compliant with the SD Host Controller Standard Specification 3.0 is available on Edison pins 44, 58, 60, 62, 64, 66, 68, and 70. The SD memory interface has a maximum 50-MHz clock rate and features card detection (insertion/removal) with a dedicated card-detection signal. The interface does require an external level-shifter chip to support 2.85-V SD memory cards. The dev board has an installed SD card holder that allows you to insert a standard SD memory card to expand the available memory. The card must be mounted using the appropriate Linux command prior to its being recognized by the system.

Sparkfun also has a micro SD card module available as part of their stackable kit. This module is shown in Figure 8-15.

I would recommend that you not try to develop your own SD card interface, but simply use either the one on the dev board or the Sparkfun module. While it is not too hard to come up with your own version, it just seems that your time and effort could be better utilized in other more ingenious project-development matters than in a reinventing-the-wheel type effort.

Clock Outputs

The clock signals are the last of the parallel interfaces to be discussed. The Edison provides two clock pulse signal waveforms, as described in Table 8-6.

Figure 8-15 *Sparkfun micro SD card module.*

The two clock outputs are available to experimenters by using the mini-breakout board and not the Arduino dev board. I would also recommend using a digital divider circuit with the high-frequency clock signal as a source to generate frequencies less than the maximum rate. Figure 8-16 shows the high-frequency clock waveform that was measured using my USB oscilloscope between pins J17-13 and J19-3 on the mini-breakout board. Note, that no programs have to be running or commands issued, as this signal is always present whenever the Edison is powered on.

These clocks are an ideal source for constant frequency signals, without requiring a committed GPIO pin or any driver software. The waveform shown in the figure is extremely accurate in the frequency, showing a 52.08 ns period, which is equivalent to 19.20 MHz. The waveform is somewhat distorted, resembling more of a "sharktooth" pattern than a square wave. I would, thus, recommend that you install a buffer chip between the Edison clock output and the external circuit to protect the Edison, sharpen the waveform transitions, and increase the peak clock waveform voltage to either 3.3 or 5 V DC.

Description	Edison pin #	Frequency	Accuracy (ppm)	Remarks
High frequency	67	19.2 MHz	+/−100	1.8-V output, 50% duty cycle
Sleep	7	32.768 KHz	+/−100	1.8-V output, 50% duty cycle

Table 8-6 *Edison Clock Outputs*

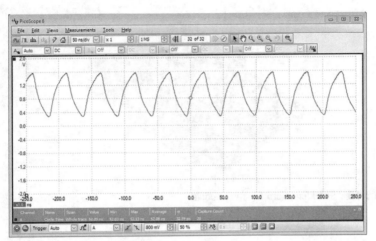

Figure 8-16 *High frequency clock signal.*

This section on clocks concludes the hardware interface discussion. The next book chapter will focus on software applications in lieu of the hardware side that I covered in this one.

Summary

This chapter described all the hardware interfaces that are available with the Edison. There are both serial and parallel interfaces, which provide different functionalities to increase the Edison's utility. I provided several demonstration projects for the serial interfaces, which should help you with your own projects that might contain similar interfaces.

9

Web Server and Database

In this chapter, I will show you how to greatly expand the Edison's utility by both incorporating an easily modified web server and connecting to a personal database. This setup will make possible all sorts of projects that can collect sensor data and also make that data remotely accessible using a remote web browser.

LAMP

LAMP is short for:

- Linux
- Apache
- MySQL
- Perl/PHP/Python

This LAMP installation requires the Debian Linux distribution to be installed in the Edison. I did not try to install LAMP with the Poky distribution because that would entail a completely new build using the Yocto framework.

The remaining three elements will be the subject for the rest of this chapter. The next step is installing an Apache web server. PHP will also be installed in this step, as it is traditionally closely allied to the Apache software. I will not be using the Perl scripting language; however, I will use Python to create a database connection to retrieve data as requested from a remote web client.

Apache Web Server and the PHP Scripting Language

The Apache web server is by far the most popular open-source web server in existence. It is very mature, having been created and updated for almost 18 years as of the time of this writing. Its formal name is Apache HTTP Server, and its logical name, as far as the Linux OS is concerned, is httpd. The "d" in the logical name stands for daemon, which is a background task in the OS lexicon. The latest version of Apache is 2.4.7, which is why it will be referred to as "apache2" during the install process.

Currently, Apache serves well over 100 million websites worldwide, which accounts for about 55 percent of all active Internet websites. This makes this software the most popular web server ever used. By some accounts, Apache has been the prime reason why the World Wide Web has been so popular.

The PHP web scripting language will also be installed in this step, as it is closely integrated with the Apache web server software. Just a bit of history regarding PHP would be helpful in understanding what it is and how it relates to Apache. PHP is a web development scripting language that is hosted on the web server that it supports. PHP originally stood for "Personal Home Page" but that has been superseded by a fancier phrase, "PHP:Hypertext Preprocessor," which is a humorous backronym. In any case, PHP is mainly used to create dynamic web pages that the web server generates in real time based upon client requests. I will use PHP in the "Hello World" project.

Well, enough with the Apache and PHP promotions, its time to install them on the Edison.

But first we must ensure that the Debian software is the most current with all modifications, new versions, and patches installed. Enter the following commands at the `root` level:

```
apt-get update
apt-get upgrade
```

It is now time to install Apache and PHP, assuming all the updates and upgrades have been completed. Type the following at the command-line prompt:

```
apt-get install apache2 php5 libapache2-mod-php5
```

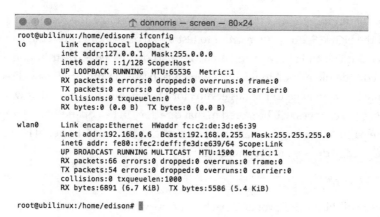

Figure 9-1 *ifconfig command results.*

The software install takes only several minutes and uses a little over 24MB. It is also prudent to restart the web server after the install. You do this by entering the following at the command-line prompt:

```
service apache2 restart
```

You will next need to use a remote web browser to confirm that Apache installed correctly. To do this, you will need to determine the IP address that the Edison was issued when it logged on to your network. The Edison uses only the wlan0 wireless adapter logical name. Check the IP address associated with wlan0 after you enter:

```
ifconfig
```

Figure 9-1 is a screenshot of the ifconfig command results. As you can see, the local IP address shown in the wlan0 section is 192.168.0.6. Your results should differ somewhat depending on the router you use and the number of devices connected to your network.

Figure 9-2 shows the web page appearing in the remote web browser, confirming that the Apache web server was working.

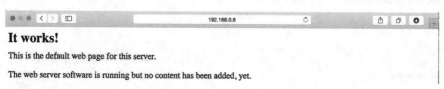

Figure 9-2 *Initial web page served by Apache.*

Not too impressive, but it is a start. You should replace the test file that produced the Figure 9-2 display in order to test if PHP is working. I use the nano editor, as it is quick and effective and is already part of the Debian distribution. Using nano, edit the file named index.html, which is located in the /var/www directory. I usually like to change to the target directory before I edit because that way I am sure that any changes will be stored in that directory in lieu of another if I forget to add the path. Assuming you start in your home directory, which should be /root, all you need to do is issue the command

```
cd /var/www
```

and you will be in the proper Apache directory that contains the default web page file, index.html. Start the nano editor with this command:

```
nano index.html
```

I deleted everything in the file and added the following:

```
<html>
<head>
<title>PHP Test</title>
</head>
<?php echo <p>Hello World</p>
</body>
</html>
```

After you have entered the code, save it by simultaneously pressing the control and "o" keys. Then simply press the enter key when the nano editor asks if you wish to save it as index.html. Press the control and "x" keys together to exit the nano editor.

Figure 9-3 shows the web page displayed when visiting this web page using a remote web browser, as was done above. It clearly shows that the PHP software is working with no issues evident. I did one more PHP test using slightly more complex code as compared to the first one. However, for this test, I created a new file named hello.php and stored it in the same directory as the index.html file. Below is the code listing for this new file.

Hello World

Figure 9-3   *PHP test web page.*

```
<?php print  <<< EOT
<!doctype html>
<html lang="en">
<head>
<meta charset="UTF-8">
<title>PHP Test</title>
</head>
<body>
<h1>PHP TEST WAS SUCCESSFUL</h1>

<p>This test confirms PHP is working on the
web server</p>
<p>The Apache web server and PHP are good to go</p>
</body>
</html>
EOT;
?>
```

When you visit this web page, you must specifically request the file named hello.php, or else the default index.html file will be displayed. Enter the following in the browser URL line:

```
http://192.168.0.6/hello.php
```

Obviously, substitute your own IP address for the one shown in the above line. Figure 9-4 shows the results for this operation.

The remaining step in creating a complete LAMP project is to install and test the MySQL database, which I do in the next section.

MySQL Database Installation

The fourth and final component for our installation is the MySQL database software. MySQL is an open-source, full-featured relational database that is an essential part of any meaningful Internet of Things (IoT) project. It really makes

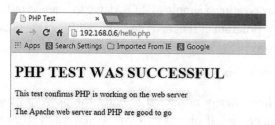

Figure 9-4   *hello.php test results.*

no sense to acquire data unless you store it for some purposeful action. I realize that real-time data can be pushed onto the web, but somewhere, somehow it should be stored for later retrieval and/or analysis. Storing it at the generation site makes sense to me, if you do not have a 100-percent reliable and continuous web connection.

This is an interesting and continually debated topic as to whether or not sensor-driven websites should have any organic, or built-in, data storage. My take on it is that MySQL doesn't cost anything other than your time to install and maintain it, and it does not really occupy a lot of memory, so why not increase your options and website reliability by using it.

MySQL has been around for a fair number of years, having been created in 1995 about the same time Apache and PHP came into existence. The installation is very straightforward and can be commenced by typing the following command:

```
apt-get install mysql-server mysql-client php5-mysql
```

It will take about eight minutes and require about 95MB of file space to completely install both the MySQL server and the client programs along with some PHP support programs. Near the end of the installation, you will be prompted to enter a `root` password for the MySQL server. I highly recommend you do so because it adds another level of security to the LAMP project.

Testing the MySQL installation is somewhat involved because you need to create and partially populate a database to evaluate whether or not the installation was successful. Before I show you these steps, I wish to point out that presenting all the essential elements and background for a relational database would take a separate book all to itself. I will show you the necessary commands on how to create and populate a MySQL database without providing an in-depth explanation of the theory. I would urge all interested readers to either take a formal course in database technology or read one of the many fine books that explain relational databases to gain a comprehensive education.

I will begin by creating a simple MySQL database named test. You must first start the MySQL program at the `root` level by entering the following at a command-line prompt:

```
mysql  -u root -p
```

You will be immediately prompted to enter the password that you entered during the install process. If all goes well, you should see the cursor waiting at the "`mysql >`" prompt. You are now in a MySQL shell program that receives, interprets, and finally executes your commands.

I do have to provide some basic background on how a database is structured before demonstrating how to store data into it. A relational database shares some resemblance to a spreadsheet, which most readers probably use. It is arranged in columns called fields and rows named records. The rows are indivisible, unlike the spreadsheet, meaning a row or record is treated as a single unit. The fields in each record are individually addressable but can exist outside of the record. Also, records must be unique; there cannot be any duplicate records in a formally constructed, or normalized, database to use the appropriate database terminology. I have created a simple example below to help clarify these concepts.

Suppose you have created a distributed, temperature-monitoring system from which you wish to log a series of temperature sensor readings along with the date, time, and sensor location. A sample log entry would consist of the following:

Date	Time	Location	Temperature

This would constitute a sample record that would be part of the table structure that, in part, makes up a database. All databases have one or more tables and each table has multiple records. tables must be created and named prior to use. Also, the record elements, such as Date and Time, are table fields, and as such, must have names to allow data to be stored in their respective fields. For this example, I decided to use the names as specified in this book's Table 9-1.

NOTE   *I will be very careful in trying to distinguish between my book's Tables and database tables, using capital "Tables" for the book and lowercase "tables" when referring to the database variety.*

Name	Description	Data Type
test	Database name	N/A
tempData	table Name	N/A
tdate	Date field	DATE
ttime	Time field	TIME
tloc	Location field	TEXT
temp	Temperature field	NUMERIC

Table 9-1   *Example Database Names and Types*

Table 9-1 is all that's needed to start creating the database, table, and fields. I have created the database schema using database parlance. Admittedly, it is very simple, but it's all that's needed for this example. The following command should be entered at the `mysql` prompt to create the database:

```
CREATE DATABASE test;
```

The semicolon is very important for MySQL commands because it indicates the end of the command sequence. Neglecting to add it will cause an error or prevent the desired command from being executed. I also want to mention that commands can extend over several lines. You will notice that you are in line continuation mode when the `mysql>` prompt changes to an indented → symbol. Just remember to always end the command sequence with a semicolon or a "\g" (backslash g).

You should also have noted that the "CREATE DATABASE" command is capitalized. This format signifies that it is an SQL command. Traditionally, all SQL commands, whether issued in the command shell or done programmatically, are capitalized. The MySQL program doesn't distinguish between capitalized and lowercase commands, but other SQL programs are more strict. I will follow the standard format and capitalize any SQL command.

The next command instructs MySQL to use the database that was just created:

```
USE test;
```

There could be several databases that were already created and in the MySQL "library." It is important that the desired database be used, which is the reason for this command.

Next, the table that will store the temperature data needs to be created. The table and all the associated fields for that table are created using this single command:

```
CREATE TABLE tempData (tdate DATE, ttime TIME, tloc
TEXT, temp NUMERIC);
```

Use the following command if you wish to check if the table and fields were properly created:

```
SHOW COLUMNS FROM tempData;
```

Figure 9-5 is a screenshot of the results of this command. This figure shows not only the field names and data types associated with the fields but

```
mysql> SHOW COLUMNS FROM tempData;
+--------+---------------+------+-----+---------+-------+
| Field  | Type          | Null | Key | Default | Extra |
+--------+---------------+------+-----+---------+-------+
| tdate  | date          | YES  |     | NULL    |       |
| ttime  | time          | YES  |     | NULL    |       |
| tloc   | text          | YES  |     | NULL    |       |
| temp   | decimal(10,0) | YES  |     | NULL    |       |
+--------+---------------+------+-----+---------+-------+
4 rows in set (0.00 sec)

mysql> █
```

Figure 9-5 *SHOW COLUMNS command result.*

also whether or not Null values will be accepted for a field entry when a record is inputted. There must be a YES in the field's NULL column to allow for a missing data value, or else the record will not be entered into the table. Whether or not to accept partial data records is an important decision to make. This decision should be made on a case-by-case basis, depending on the nature of the data logging that is desired. The figure also indicates whether or not a default value should be entered in the case where there is no actual value present in the input record. Usually this is not needed, as that situation is more easily handled programmatically.

I next used INSERT statements to manually enter three records into the temp-Data table. The following is an example of one of these INSERT statements

```
INSERT INTO tempData (tdate, ttime, tloc, temp) VALUES
(DATE('2015-02-25'), TIME('09:30:00'), 'Garage', 22);
```

Observing the length of this statement makes one appreciate the usefulness of the line continuation prompts that automatically appear in the MySQL shell. The statement really isn't all that complex when you break it into its major parts as I have done below:

- INSERT INTO—The SQL command to insert what follows into a record for the specified table.

- tempData—The specified table. Remember that I earlier issued the command "USE test;" to specify the database that contains the tempData table.

- (tdate, ttime, tloc, temp)—The fields that you wish to provide data for, or "populate." You may skip one or more fields if Null values are permitted, as discussed above.

- VALUES—The SQL expression that indicates that the data follows. It is important to provide a matching number of data values and types to those specified after the table's name. Don't enter TEXT if you specified a NUMERIC type.

- DATE('2015-02-25')—This value entry takes advantage of the built-in MySQL DATE function, which converts a string into a DATE data type. There are two acceptable string formats, YYYY-MM-DD and YY-MM-DD.

- TIME('09:30:00')—This value entry takes advantage of the built-in MySQL TIME function, which converts a string into a TIME data type. The acceptable string format is HH:MM:SS.

- 'Garage'—A string indicating the data sensor's location.

- 22—The numeric temperature value in °C.

I entered two additional records after this initial one, and I took advantage of a real time-saver for MySQL shell data entry. Just press the up-arrow key to retrieve the last command entered, which allows you to edit it with the new data. You can repeatedly press either the up- or down-arrow keys and quickly scan the total MySQL command-history buffer. Highly recommended as a time-saving tip. In fact, this also works for the Linux command-line buffer. I use these keys all the time.

Figure 9-6 shows the three data records that were manually entered using the INSERT INTO command. I used the following SQL command to display the three new records:

```
SELECT * FROM tempData;
```

But take heart. All the book's projects use a program to enter data, and you will not have to enter any data manually unless you wish to edit a record.

```
mysql> SELECT * FROM tempData;
+------------+----------+---------+------+
| tdate      | ttime    | tloc    | temp |
+------------+----------+---------+------+
| 2015-02-25 | 09:30:00 | Garage  |   22 |
| 2015-02-25 | 09:30:10 | Case    |   26 |
| 2015-02-25 | 09:30:20 | Outside |   24 |
+------------+----------+---------+------+
3 rows in set (0.00 sec)
```

Figure 9-6 *tempData table with three records.*

```
mysql> DELETE FROM tempData;
Query OK, 3 rows affected (0.06 sec)
```

Figure 9-7   *tempData table contents after the* DELETE FROM *command.*

I don't want to leave this section without showing you how to delete data. Simply use the following command to delete all the records from the tempData table:

```
DELETE FROM tempData;
```

Be cautious, as this is a powerful command that will irretrievably delete all the data records. There is no recourse unless you have separately backed up the data. Also, realize that the tempData table schema is not affected by this command. All the field names and associated specifications remain intact. Figure 9-7 shows the results of removing all the data from the tempData table.

If you wish to remove only certain select records, you can use what is known as a "WHERE" clause. For the above example, suppose I wanted to delete only the records containing the word Case in the tloc field, I would rewrite the DELETE command to be as follows:

```
DELETE FROM tempData WHERE tloc='Case';
```

I applied this command to the tempData table after I restored the three records, using the buffer key hint I gave above. Figure 9-8 shows the result in which the record containing the tloc field value of "Case" was deleted. Actually, all records that contained "Case" in the tloc field would have been deleted, if there had been multiple records with that tloc field value.

The WHERE clause is a very valuable tool that can be applied to most of the MySQL commands. Applying selective operations to table records is a key

```
mysql> DELETE FROM tempData WHERE tloc='Case';
Query OK, 1 row affected (0.06 sec)

mysql> SELECT * FROM tempData;
+------------+----------+---------+------+
| tdate      | ttime    | tloc    | temp |
+------------+----------+---------+------+
| 2015-02-25 | 09:30:00 | Garage  |   22 |
| 2015-02-25 | 09:30:20 | Outside |   24 |
+------------+----------+---------+------+
2 rows in set (0.00 sec)
```

Figure 9-8   *tempData table after the* DELETE FROM *with a* WHERE *clause command.*

database manipulative element that you should find very handy, especially with large datasets.

Adding a New User to a MySQL Database

Up to now I have been logging into MySQL as "root" using the command:

```
mysql  -u root -p
```

This is not a very good practice, especially if you desire to have the database be made available to others. I will now show how easy it is to create new users and have them attached to specific databases. This is a very nice way to allow different users to use only the databases they need to access. Now you have the ability to maintain many databases and restrict access to each specific database only to those users who are authorized to use it.

You must first login as root to have the privileges to create new users. Use the login command shown above. Suppose there is a new user named tester1 who needs access to the test database. Enter the following command to create the user in MySQL:

```
CREATE USER 'tester1' IDENTIFIED BY 'password';
```

Of course, I have purposefully used the worst password known: "password," but this was done only to keep it simple. You must next grant this new user privileges in order to perform operations on the target database. You do this with the following command:

```
GRANT ALL PRIVILEGES ON test.* TO 'tester1';
```

Notice the command element "test.*," which tells MySQL that the user tester1 has access to all components of only the test database and to no others. However, the user can perform all MySQL commands without restriction on that database.

Next you should issue the command:

```
FLUSH PRIVILEGES;
```

This forces the MySQL grant table, which holds all user privileges, to reload, thus, saving the new privileges that you just established for the user tester1. Next quit MySQL so you may login as tester1.

```
quit;
```

```
mysql> USE test1;
ERROR 1044 (42000): Access denied for user 'tester1'@'%' to database 'test1'
mysql> ▌
```

Figure 9-9   *Tester1 error message.*

Now login to MySQL as tester1 with this command:

```
mysql -u tester1 -p
```

You will now be asked for tester1's password, which I just gave you. You will now be at the "mysql >" prompt from which you can enter all the commands as was done for user "root."

Being a naturally curious individual, I decided to confirm whether or not the tester1 user was confined to the test database only and had no access to any other one. I quit the MySQL application and logged in as root. This time I created a null database named test1. A null database, as the name implies, has no content. I then quit the application and logged in again as the tester1 user. I then tried the "USE test1;" command to try to switch from the test database to the newly instantiated test1 database. Figure 9-9 shows the resultant error message confirming that MySQL will not allow a non-root user into any database other than the one to which the user is registered.

This last error check ends this section. I will next demonstrate how to remotely connect to a database as well as discuss many other interesting and related topics.

Python Database Connection

An open-source Python package named python-mysqldb contains all the libraries necessary to establish connectivity between a Python program (script) and a MySQL database. You need to install the package by entering this command:

```
apt-get install python-mysqldb
```

Next you should create a test program named mysqlTest.py to confirm that the Python to MySQL connection works. Enter the following to start the nano editor and then enter the code, which follows the command.

```
nano mysqlTest.py
```

NOTE   *I used the test database created earlier. It was quick and convenient and already had some sample data in the* tempData *table. I also logged in as the* tester1 *user that I had previously added to the database.*

```
#!/usr/bin/python
# mysqlTest.py
# D. J. Norris
# The code is in the public domain

import MySQLdb

# this creates the basic connection object
db = MySQLdb.connect( host='localhost', user='tester1',passwd=
'password',db='test')

# a cursor object is required to execute SQL commands on the database
cur = db.cursor()

# this SQL command retrieves all the records from the tempData table
cur.execute('SELECT * FROM tempData')

# print the field headers
print "DATE                TIME           LOCATION     TEMP (C)"

# display all the data record after record
for row in cur.fetchall():
    print row[0], "    ", row[1], "     ", row[2], "     ", row[3]
```

Execute the mysqlTest.py program by entering the following at the command line:

```
python mysqlTest.py
```

Figure 9-10 shows the resulting display after the program is run. All the records that were previously manually entered are displayed, one record at a time.

Running this program has demonstrated how relatively easy it is to instantiate a MySQL database connection to a Python application and retrieve the data records. I next would like to describe how to build an automatic, multichannel temperature measurement system before I show you how to programmatically insert data into a MySQL database.

```
● ● ●              ⌂ donnorris — screen — 80×24
root@ubilinux:~# python mysqlTest.py
DATE            TIME        LOCATION      TEMP (C)
2015-02-25      9:30:20     Outside       24
2015-02-25      9:30:20     Outside       24
2015-02-25      9:30:20     Case      26
2015-02-25      9:30:30     Garage        22
2015-02-25      9:30:40     Case      27
2015-02-25      9:30:40     Garage        23
root@ubilinux:~#
```

Figure 9-10 *mysqlTest program results.*

Home Temperature Measurement System

Figure 9-11 shows a block diagram of the system with three sensors that are connected using wires to an interface block that, in turn, connects to the dev board, using the bit-banged software interface described in the previous chapter. I set up the system test in two phases, with the first using a single sensor and the second using all three sensors.

I used the Analog Devices TMP36 temperature sensor because it is very inexpensive, solid-state, and easy to hook up. This sensor is described in the following section.

TMP36 Temperature Sensor

The basic temperature sensor I will use in this project is an Analog Devices model TMP36 shown in Figure 9-12. It is housed in a standard TO-92 plastic form factor that is also common to most transistors. The TMP36 is far more complex than a simple transistor in that it contains circuits to both sense ambient temperature

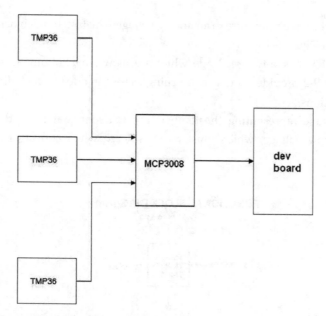

Figure 9-11   *Temperature monitoring system block diagram.*

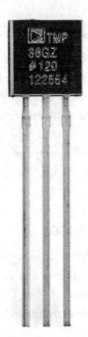

Figure 9-12   *Analog Devices model TMP36 temperature sensor.*

and convert that temperature to an analog voltage. The functional block diagram is shown in Figure 9-13.

The TMP36 has only three leads, which are shown in a bottom view in Figure 9-14. Table 9-2 provides details concerning these three leads, including important limitations.

The voltage representing the temperature is dependent upon the TMP36 power supply voltage, which must be considered when converting the V_{OUT}

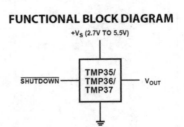

Figure 9-13   *Model TMP36 functional block diagram.*

PIN 1, +V$_S$; PIN 2, V$_{OUT}$; PIN 3, GND

Figure 9-14 *TMP36 bottom view showing external leads.*

voltage to the equivalent real-world temperature. I do account for this in the software that converts the V$_{OUT}$ voltage to an actual temperature. Figure 9-15 is a graph of the V$_{OUT}$ voltage versus temperature, using a 3-V supply voltage

The actual temperature measurement range for the TMP36 is −40 to +125°C, with a typical accuracy of +/−2°C and a 0.5°C linearity. All-in-all, not too shabby specifications, considering that the cost of the TMP36 is typically less than $2 USD. The TMP36 range, accuracy, and linearity are well suited for a home temperature monitoring system.

Initial Test

Initial testing involves both creating a hardware circuit using a single sensor and establishing the proper Python software environment. I will be using the bit-banged software that I discussed in the previous chapter to interface the TMP36 sensors to an MCP3008 ADC. The dev board will, in turn, accept the sampled voltage created by the ADC and convert the measurements into real temperatures.

Hardware Setup

I will first discuss the hardware circuit, since that is relatively straightforward. Figure 9-16 shows the test schematic for the dev board, MCP3008, and TMP36. I connected the TMP36 V$_{OUT}$ lead to the MCP3008 channel 0 input, which is pin 1.

Pin Number	Description	Remarks
1	+V$_s$	Supply voltage. Ranges from 2.7 to 5.5 V. This project uses 3.3 V.
2	V$_{OUT}$	The analog voltage representing the temperature. The maximum voltage depends upon the supply voltage.
3	GND	Common reference used by both the supply and V$_{OUT}$ pins.

Table 9-2 *TMP36 Pin Details*

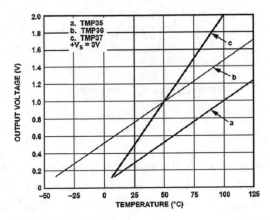

Figure 9-15   *Graph of V_{OUT} voltage versus temperature for $+V_S = 3$ V.*

I mounted the TMP36 and MCP3008 on a solderless breadboard, which, in turn, was mounted on a Lexan plate along with the dev board, as shown in Figure 9-17. The plate will eventually be placed in a plastic case when the multiple sensor system is built.

On the right side of the breadboard, you can see a TMP36 sensor connected with three jumper wires to the MCP3008. The hardware setup should proceed very quickly, after which you can proceed to the next portion of the test: the installation of the driver software.

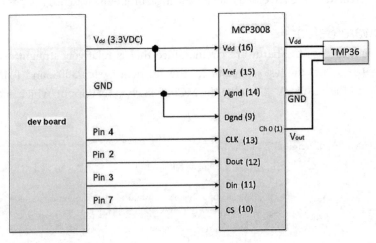

Figure 9-16   *Test schematic.*

Figure 9-17 *Physical test setup.*

Software Setup

I used a modified version of the TestSPI_ADC.py program, which I discussed in the previous chapter. The following test program displays a continuous stream of temperature values generated by the TMP36 sensor. The program is named SingleSensorTest.py.

```
#!/usr/bin/python

# SingleSensorTest.py
#   D. J. Norris 2/2015
# This code is in the public domain

import time
import mraa as m

# define a function to read the MCP3008 ADC value
def adc(chan, clock, mosi, miso, cs):
    if((chan < 0) or (chan > 7)):
        return -1
    cs.write(True)
    clock.write(False)
    cs.write(False)
    cmd = chan
    cmd |= 0x18
    cmd <<= 3

    for i in range(5):
        if(cmd & 0x80):
            mosi.write(True)
        else:
            mosi.write(False)
```

```
        cmd <<= 1
        clock.write(True)
        clock.write(False)
    result = 0
    for i in range(12):
        clock.write(True)
        clock.write(False)
        result <<= 1
        if(miso.read()):
            result |= 0x1
    cs.write(True)
    result >>= 1
    return result

#These pin definitions are set to work with the test circuit
SPICLK = m.Gpio(4)
SPIMISO = m.Gpio(2)
SPIMOSI = m.Gpio(3)
SPICS = m.Gpio(7)

# Set pins for direction
SPICLK.dir(m.DIR_OUT)
SPIMISO.dir(m.DIR_IN)
SPIMOSI.dir(m.DIR_OUT)
SPICS.dir(m.DIR_OUT)

# use channel 0 as a test input
channel = 0

# define a function that converts data to voltage levels
# round to a specified number of decimal places
def ConvertVolts(data, places):
    volts = (data * 3.3) / 1023
    volts = round(volts, places)
    return volts

# define a function to calculate temperature from TMP36 data
# round to a specified number of decimal places
def ConvertTemp(data, places):


    #ADC Value       Temp (°C)        Volts
    #    0            -50             0.00
    #    78           -25             0.25
    #    155            0             0.50
    #    233           25             0.75
    #    310           50             1.00
    #    465          100             1.50
    #    775          200             2.50
    #   1023          280             3.30

    temp = ((data * 346)/1023) - 50   # constant 346 is empirically set
    temp = round(temp, places)
    return temp
```

```
#define temp channel
channel = 0

#define time between readings
delay = 5

#print column headers
print "temp_level            temp_volts         temp"

#main loop
while True:
    #read an ADC value
    temp_value = adc(channel, SPICLK, SPIMOSI, SPIMISO, SPICS)
    temp_volts = ConvertVolts(temp_value, 2)
    temp = ConvertTemp(temp_level, 2)

    #display results
    print "-------------------------------------"
    print temp_level, "          ", temp_volts, "        ", temp

    #delay before taking next measurement
    time.sleep(delay)
```

Run the above program by entering:

```
python SingleSensorTest.py
```

Figure 9-18 is screenshot of a portion of the program output with the TMP36 sensor measuring the temperature within the case.

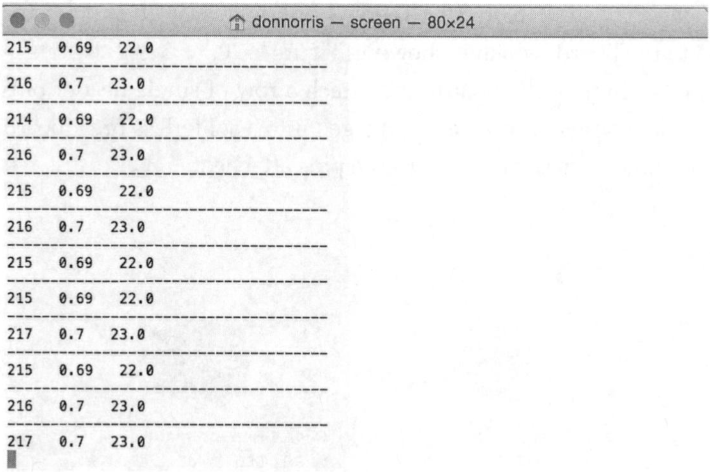

Figure 9-18 *Initial test results.*

I put a comment in the code listing explaining that I needed to adjust a constant in the temperature conversion function from the original 330 value to a 346 value. I did this because the MCP3008 chip was reporting a voltage of 0.70 V, while the actual voltage, as measured with a digital voltmeter, was approximately 0.74 V. This caused the temperature to be underreported by approximately 4°C. I also used a non-contact, precision infrared temperature meter to measure the real temperature. By adjusting the constant, I forced the function to calculate the correct temperature. I do not know why the MCP3008 chip was not accurately converting the TMP36 V_{OUT}, but I did confirm that the error was linear and constant, thus could easily be corrected in the conversion formula.

Multiple Sensor System

It is now time to build a three-channel system now that the single-channel test has proved that the ADC, sensor, and supporting software all function as expected. I used a wired system because it simplified the design, allowed the focus to be on the sensors, ADC, and software, and made the connections quick, easy, and flexible. Those were the reasons I chose RJ45 cables and connectors for the wiring component. I did find an RJ45 breakout board from Sparkfun, which along with a companion connector made the cable and sensor connections quite easy and convenient. The connector and breakout board are shown in Figure 9-19 along with the Sparkfun model numbers.

You must first push the connector onto the board and then carefully solder all eight of the PCB pins. Note that there is only one way the connector can be attached to the board, which is shown in Figure 9-20.

Next, take three of the boards and attach a row of single header pins, which will allow the boards to be directly plugged into a solderless breadboard. Figure 9-21 shows one of these boards with the pins attached.

Breakout Board for RJ45
BOB-00716

RJ45 8-Pin Connector
PRT-00643

Figure 9-19 *Sparkfun RJ45 breakout board and connector.*

Figure 9-20 *RJ45 connector attached to a breakout board.*

The other three boards have a TMP36 sensor directly attached to them, as shown in Figure 9-22. Ensure you solder the sensor to the pins on the left side of the breakout board with the sensor's flat side pointing up, as shown in the figure.

You will now need the interconnecting RJ45 cables, which may be bought or made. I would suggest that you purchase them if you do not have any experience in making up this cable type. It does require a special tool along with the Cat 5 or Cat 6 cable and ready-to-assemble snap-on connectors. The cable lengths depend upon the spacing between the sensor locations and the Pi's location. I used three six-foot cables for my setup, as it was a temporary demonstration system and not a permanent one.

However, I did find that there is a tremendous difference in cable quality such that one cable would reduce the output voltage by 16 mV, while another would have hardly any effect. A 16-mV drop would cause the temperature to be sensed

Figure 9-21 *RJ45 board with attached header pins.*

Figure 9-22 *RJ45 board with an attached TMP36 sensor.*

at 2 °C less than the real temperature. The drop in voltage is likely due to the very limited output current capacity of the TMP36 sensor along with the variability in the manufacture of the cables—some with greater capacitance and inductive loading than others. I did check the TMP36's manufacturer's datasheet in which the limited current capacity was acknowledged. There is a circuit in the datasheet shown in Figure 9-23 that will boost the current level to a full-scale maximum of 2 mA, which should be more than sufficient to drive any cable attached to the sensor.

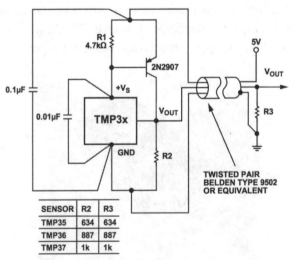

Remote, 2-Wire Boosted Output Current Temperature Sensor

Figure 9-23 *TMP36 current boost circuit.*

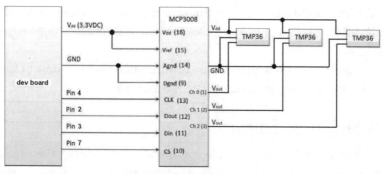

Figure 9-24 *Complete multiple sensor system schematic.*

Instead of building the boost circuit, I found it simpler just to sort through all my spare RJ45 patch cables until I found three that did not significantly affect the sensor. However, the current boost circuit is likely essential if you plan on setting up a long cable run of more than 10 feet.

The complete system schematic is shown in Figure 9-24. It is the same as the Figure 9-16 schematic with two additional sensors connected to the MCP3008 channels 1 and 2.

The physical setup with the dev board is shown in Figure 9-25, with the Lexan plate now screwed into a sturdy plastic case, which provides protection

Figure 9-25 *Physical multiple sensor system arrangement.*

from the weather. Notice how I arranged the two RJ45 connectors on the breadboard for easy hookup and cable attachment through cutouts in the side of the case.

This completes the system hardware configuration, and it is time to focus on expanding the software to accommodate two additional sensors.

Multiple Sensor Software

The software controlling the multiple sensor system is essentially the same as the single sensor version except for the two additional sensors. However, the program will be revised to display the date, time, and channel number as well as the temperature. I also do not display the temp_value or temp_volts variables in this version. All the new data will eventually be required for the database version that follows later in the chapter. In addition, I still "hardcode" the temperature channels into the program, i.e., channel 0 is sensor 0, channel 1 is sensor 1, and channel 2 is sensor 2. I recognize that this limits the programs flexibility, but in the interests of simplicity, I don't believe it is too much of a compromise. I also doubled the delay time to reduce the data flow a bit.

The new program is named MultipleSensorTest.py.

```python
#!/usr/bin/python
# MultipleSensorTest.py
#  D. J. Norris 2/2015
# This code is in the public domain

import time
import mraa as m

# define a function to read the MCP3008 ADC value
def adc(chan, clock, mosi, miso, cs):
    if((chan < 0) or (chan > 7)):
        return -1
    cs.write(True)
    clock.write(False)
    cs.write(False)
    cmd = chan
    cmd |= 0x18
    cmd <<= 3

    for i in range(5):
        if(cmd & 0x80):
            mosi.write(True)
        else:
            mosi.write(False)
        cmd <<= 1
        clock.write(True)
```

```
        clock.write(False)
    result = 0
    for i in range(12):
        clock.write(True)
        clock.write(False)
        result <<= 1
        if(miso.read()):
            result |= 0x1
    cs.write(True)
    result >>= 1
    return result

#These pin definitions are set to work with the test circuit
SPICLK = m.Gpio(4)
SPIMISO = m.Gpio(2)
SPIMOSI = m.Gpio(3)
SPICS = m.Gpio(7)


# Set pins for direction
SPICLK.dir(m.DIR_OUT)
SPIMISO.dir(m.DIR_IN)
SPIMOSI.dir(m.DIR_OUT)
SPICS.dir(m.DIR_OUT)

# use channels 0, 1, and 2 as a test inputs
channel0 = 0
channel1 = 1
channel2 = 2

# define a function that converts data to voltage levels
# round to a specified number of decimal places
def ConvertVolts(data, places):
    volts = (data * 3.3) / 1023
    volts = round(volts, places)
    return volts

# define a function to calculate temperature from TMP36 data
# round to a specified number of decimal places
def ConvertTemp(data, places):

    #ADC Value        Temp (°C)        Volts
    #    0             -50             0.00
    #    78            -25             0.25
    #    155            0              0.50
    #    233            25             0.75
    #    310            50             1.00
    #    465           100             1.50
    #    775           200             2.50
    #   1023           280             3.30

    temp = ((data * 330)/1023) - 50
    temp = round(temp, places)
    return temp
```

```
#define temp channel
channel0 = 0
channel1 = 1
channel2 = 2

#define time between readings
delay = 10

#print column headers
print "Date          Time          Chan No      Temp"

#main loop
while True:

    # get the current system date in the format dd/mm/yyyy
    sample_date = time.strftime("%d/%m/%Y")

    # get the current system time in the 24 hour format hh:mm:ss
    sample_time = time.strftime("%H:%M:%S")
    #you change the format to 12 hour by substituting %I for %H

    #read all three temp channels
    temp_value0 = adc(channel0, SPICLK, SPIMOSI, SPIMISO, SPICS)
    temp_value1 = adc(channel1, SPICLK, SPIMOSI, SPIMISO, SPICS)
    temp_value2 = adc(channel2, SPICLK, SPIMOSI, SPIMISO, SPICS)

    temp_volts0 = ConvertVolts(temp_value0, 2)
    temp_volts1 = ConvertVolts(temp_value1, 2)
    temp_volts2 = ConvertVolts(temp_value2, 2)

    temp0 = ConvertTemp(temp_volts0, 2)
    temp1 = ConvertTemp(temp_volts1, 2)
    temp2 = ConvertTemp(temp_volts2, 2)

    #display results
    print "-------------------------------------------------------------
_"
    print sample_date, "     ", sample_time, "     ", channel0, "
", temp0
    print sample_date, "     ", sample_time, "     ", channel1, "
", temp1
    print sample_date, "     ", sample_time, "     ", channel2, "
", temp2

    # delay before taking next measurement
    time.sleep(delay)
```

A sample program display is shown in Figure 9-26.

I also wanted to point out that I deliberately chose not to use iteration to sample and display all the sensors, as there were only three sensors and the memory saved was not as important to me as the program efficiency gained by "unrolling"

```
       donnorris — screen — 80×31
27/02/2015   00:20:54   0   22.0
27/02/2015   00:20:54   1   20.0
27/02/2015   00:20:54   2   23.0
------------------------------------------------------
27/02/2015   00:21:04   0   22.0
27/02/2015   00:21:04   1   20.0
27/02/2015   00:21:04   2   23.0
------------------------------------------------------
27/02/2015   00:21:14   0   22.0
27/02/2015   00:21:14   1   20.0
27/02/2015   00:21:14   2   25.0
------------------------------------------------------
27/02/2015   00:21:24   0   22.0
27/02/2015   00:21:24   1   20.0
27/02/2015   00:21:24   2   24.0
------------------------------------------------------
27/02/2015   00:21:34   0   22.0
27/02/2015   00:21:34   1   21.0
27/02/2015   00:21:34   2   24.0
------------------------------------------------------
27/02/2015   00:21:44   0   22.0
27/02/2015   00:21:44   1   21.0
27/02/2015   00:21:44   2   22.0
```

Figure 9-26   *Sample display from the MultipleSensorTest.py program.*

the loops. I would definitely use loops if five or more sensors were utilized because the program would otherwise become quite large and tedious to enter.

I will now change focus slightly and discuss how to create a database to store the temperature data for eventual retrieval using a web browser

Temperature Database

I will now need to create a new database to store the temperature data generated by the three-sensor system. The test database structure will be used as a template for the new database with some modifications. I will also set up a new user for this database because it would be a serious security issue to allow root access to a database that is also accessible via a web browser.

The new database is named HomeTempSystem and will have two tables in it named sensorTemp and channelLocation. The channelLocation table will enable a convenient method of describing a sensor's location and to change it as needed. The schema, or structure, for the sensorTemp table is shown in Table 9-3

The "id" is a new addition in the field listing as compared to the original *test* database, and it is described as a *primary key*. This is an important designation because there cannot be any duplicate records contained in any relational database table. The id field is simply an integer that is automatically incremented

Name	Description	Data Type
HomeTempSystem	Database name	N/A
sensorTemp	table name	N/A
id	Primary key	AUTO_INCREMENT
tdate	Date field	DATE
ttime	Time field	TIME
tchan	Channel number	NUMERIC
ttemp	Temperature field	NUMERIC

Table 9-3 *sensorTemp table Structure*

every time a new record is added. I will not explicitly use the id field in this project, but it is a good security feature that ensures that only unique records are inserted into the table. However, I will use a combination of the tdate, ttime, and tchan fields to retrieve any desired temperature data. Incidentally, auto-incrementing keys are a very efficient and fast means to retrieve large data sets from a database in lieu of a combination of field lookups.

The channelLocation table's schema is detailed in Table 9-4. It is much simpler than the previous one, as it has only two fields of which the tchan field serves as the primary key and the tloc field is used to store the text data describing the location of the particular sensor.

To create the database and tables, you must first start the MySQL program by entering:

```
mysql -u root -p
```

Next enter the password you created when you first installed MySQL, as discussed in the previous chapter.

Now create an empty database named HomeTempSystem by entering:

```
CREATE DATABASE HomeTempSystem;
```

Name	Description	Data Type
HomeTempSystem	Database name	N/A
channelLocation	table name	N/A
tchan	Primary key	NUMERIC
tloc	Sensor location	TEXT

Table 9-4 *channelLocation table Structure*

Next switch over to the new database:

```
USE HomeTempSystem;
```

Now create the `sensorTemp` table containing all of the fields as detailed in Table 9-3:

```
CREATE TABLE sensorTemp (id MEDIUMINT  AUTO_INCREMENT,
tdate DATE NOT NULL, ttime TIME NOT NULL, tchan
NUMERIC NOT NULL, ttemp NUMERIC NOT NULL, PRIMARY
KEY(id)) ENGINE=MyISAM;
```

All the fields have been designated as NOT NULL, which means that a proper value has to be present or else the record will not be entered into the table. It is another means to ensure that the table is not populated with garbage data.

Next create the `channelLocation` table containing the two fields as detailed in Table 9-4:

```
CREATE TABLE channelLocation(tchan NUMERIC NOT NULL,
tloc TEXT, PRIMARY KEY(tchan));
```

Note that the `tchan` field must match the field description as specified in the `sensorTemp` table. This is necessary because this field is the logical link between the two tables. It is the primary key in the `channelLocation` table and is also known as a foreign key in the `sensorTemp` table. This arrangement is very useful given that the channel numbers are repeated many times in the `sensorTemp` table but only one instance of a text channel description is needed in the `channelLocation` table because the `tchan` field links the two. This table linkage is one of the most valuable features of relational databases.

Data must be manually entered into the `channelLocation` table because it will not be programmatically inserted. This type of data is considered static and unchanging and is normally provided for descriptive purposes—in this case, just the text locations where the sensors are located. I used three INSERT statements to manually populate the table, one of which is shown below:

```
INSERT INTO channelLocation (tchan, tloc) VALUES
('0', 'Case');
```

You will readily be able to see the other two locations by using the SELECT command shown when I discuss the ViewRecords program later in the chapter.

A new user must next be created for the security reasons mentioned earlier. I named the new user "TempUser1," but it could be any name that suits your purposes or needs. This new user will have complete read and write privileges to the `HomeTempSystem` database only and to no others. Enter the following to first set up this user:

```
CREATE USER 'TempUser1' IDENTIFIED BY 'Px158qqr';
```

Note that a new password must be added for this user, as shown in the above command. You now need to associate this new user to the designated database by entering:

```
GRANT ALL PRIVILEGES ON HomeTempSystem.* TO 'TempUser1';
```

Finally, you need to execute the following command to actually set up the new user TempUser1's privileges, as discussed earlier:

```
FLUSH PRIVILEGES;
```

The last command completes the setup of the new database along with the new tables and user. You should next close the MySQL program by entering:

```
EXIT;
```

You can now optionally test the new database by following the procedures detailed earlier; however, I will now proceed to demonstrate how to programmatically populate the table using a Python connection and the data generated by the MultipleSensorTest program.

Inserting Data into a MySQL Database Using a Program

I will demonstrate how to insert data into the `HomeTempData` database by modifying the MultipleSensorTest.py program. You must have already created the database and table as well as added the new user TempUser1 with the associated password as shown above.

The program modifications consist of:

- Establishing a database connection
- Adding SQL statements to `INSERT` temperature data into the HomeTempSystem database
- Removing the console display statements
- Removing the channel number check

- Removing the data to voltage level function
- Removing the comments ADC/Temp/Volts table
- Extending the delay between measurements to 60 seconds

I also renamed the modified program SensorDatabase.py.

```python
#!/usr/bin/python
# SensorDatabase.py
#   D. J. Norris 2/2015
# This code is in the public domain

import time
import MySQLdb
import mraa as m

# define a function to read the MCP3008 ADC value
def adc(chan, clock, mosi, miso, cs):
    if((chan < 0) or (chan > 7)):
        return -1
    cs.write(True)
    clock.write(False)
    cs.write(False)
    cmd = chan
    cmd |= 0x18
    cmd <<= 3

    for i in range(5):
        if(cmd & 0x80):
            mosi.write(True)
        else:
            mosi.write(False)
        cmd <<= 1
        clock.write(True)
        clock.write(False)
    result = 0
    for i in range(12):
        clock.write(True)
        clock.write(False)
        result <<= 1
        if(miso.read()):
            result |= 0x1
    cs.write(True)
    result >>= 1
    return result

#These pin definitions are set to work with the test circuit
SPICLK = m.Gpio(4)
SPIMISO = m.Gpio(2)
SPIMOSI = m.Gpio(3)
SPICS = m.Gpio(7)

# Set pins for direction
```

```
SPICLK.dir(m.DIR_OUT)
SPIMISO.dir(m.DIR_IN)
SPIMOSI.dir(m.DIR_OUT)
SPICS.dir(m.DIR_OUT)

# setup channels
channel0 = 0
channel1 = 1
channel2 = 2

# define a function that converts data to voltage levels
# round to a specified number of decimal places
def ConvertVolts(data, places):
    volts = (data * 3.3) / 1023
    volts = round(volts, places)
    return volts

# define a function to calculate temperature from TMP36 data
# round to a specified number of decimal places
def ConvertTemp(data, places):
    temp = ((data * 346)/1023) - 50
    temp = round(temp, places)
    return temp

#define temp channel
channel0 = 0
channel1 = 1
channel2 = 2

#define unit delay, total delay time is set in program
delay = 1

#create the basic connection object
db = MySQLdb.connect(host='localhost', user='TempUser1',
passwd='Px158qqr', db='HomeTempSystem')
#a cursor object is required to execute SQL commands on the database
cur = db.cursor()
logTemp = "INSERT INTO sensorTemp (tdate, ttime, tchan, ttemp) VALUES
(%s, %s, %s, %s)"

#main loop
while True:

    # get the current system date in the format dd/mm/yyyy
    sample_date = time.strftime("%d/%m/%Y")

    # get the current system time in the 24 hour format hh:mm:ss
    sample_time = time.strftime("%H:%M:%S")
    #you change the format to 12 hour by substituting %I for %H

    #read all three temp channels
    temp_value0 = adc(channel0, SPICLK, SPIMOSI, SPIMISO, SPICS)
    temp_value1 = adc(channel1, SPICLK, SPIMOSI, SPIMISO, SPICS)
    temp_value2 = adc(channel2, SPICLK, SPIMOSI, SPIMISO, SPICS)
```

```
temp_volts0 = ConvertVolts(temp_value0, 2)
temp_volts1 = ConvertVolts(temp_value1, 2)
temp_volts2 = ConvertVolts(temp_value2, 2)

temp0 = ConvertTemp(temp_value0, 2)
temp1 = ConvertTemp(temp_value1, 2)
temp2 = ConvertTemp(temp_value2, 2)

#these statements insert all of the channel temp data into the
sensorTemp table
cur.execute(logTemp, (sample_date, sample_time, channel0, temp0))
cur.execute(logTemp, (sample_date, sample_time, channel1, temp1))
cur.execute(logTemp, (sample_date, sample_time, channel2, temp2))

# this next statement forces the data to be written into the
database
db.commit()

# countdown timer
# delay before taking next measurement
i = 60
while ( i != 0):
   time.sleep(delay)
   print i
   i -= 1
```

There is a hidden problem with the above program in that it will continue to run without any programmed way of stopping it until there is no more room for database records to be added. Such a situation would likely crash the Debian OS. However, if each record is about 100 bytes in length, I estimate it would take over six years to fill up 1 GB of memory. Therefore, I have no problem in doing a Control-C to manually interrupt the program, given this long time before calamity strikes. The worst that would happen is to corrupt one data record, which is quite acceptable during the development phase.

To run the program, first ensure that the multiple sensor system is attached and all the sensors are deployed as you desire them to be placed. Next enter the following command to start the logging of temperatures to the MySQL database

```
python SensorDatabase.py
```

You should now see a countdown timer near the command prompt indicating that the program is running. This is really just a trivial add-on whose only function is to indicate that the program is running. I wanted to avoid the situation in which the prompt simply disappeared and there was no indication of any activity. Let the program run for at least 30 minutes to build up a fair number of records before stopping it with the Control-C combination key press.

```
| 214 | 27/02/2015 | 15:29:29 |   0 |   19 |
| 215 | 27/02/2015 | 15:29:29 |   1 |   16 |
| 216 | 27/02/2015 | 15:29:29 |   2 |   16 |
| 217 | 27/02/2015 | 15:30:29 |   0 |   19 |
| 218 | 27/02/2015 | 15:30:29 |   1 |   16 |
| 219 | 27/02/2015 | 15:30:29 |   2 |   16 |
| 220 | 27/02/2015 | 15:31:29 |   0 |   18 |
| 221 | 27/02/2015 | 15:31:29 |   1 |   16 |
| 222 | 27/02/2015 | 15:31:29 |   2 |   17 |
| 223 | 27/02/2015 | 15:32:29 |   0 |   19 |
| 224 | 27/02/2015 | 15:32:29 |   1 |   16 |
| 225 | 27/02/2015 | 15:32:29 |   2 |   17 |
| 226 | 27/02/2015 | 15:33:29 |   0 |   20 |
| 227 | 27/02/2015 | 15:33:29 |   1 |   16 |
| 228 | 27/02/2015 | 15:33:29 |   2 |   17 |
| 229 | 27/02/2015 | 15:34:29 |   0 |   18 |
| 230 | 27/02/2015 | 15:34:29 |   1 |   16 |
| 231 | 27/02/2015 | 15:34:29 |   2 |   17 |
| 232 | 27/02/2015 | 15:35:29 |   0 |   18 |
| 233 | 27/02/2015 | 15:35:29 |   1 |   16 |
| 234 | 27/02/2015 | 15:35:29 |   2 |   17 |
| 235 | 27/02/2015 | 15:36:30 |   0 |   19 |
| 236 | 27/02/2015 | 15:36:30 |   1 |   15 |
| 237 | 27/02/2015 | 15:36:30 |   2 |   16 |
| 238 | 27/02/2015 | 15:37:30 |   0 |   18 |
| 239 | 27/02/2015 | 15:37:30 |   1 |   15 |
| 240 | 27/02/2015 | 15:37:30 |   2 |   17 |
| 241 | 27/02/2015 | 15:38:30 |   0 |   19 |
| 242 | 27/02/2015 | 15:38:30 |   1 |   16 |
| 243 | 27/02/2015 | 15:38:30 |   2 |   17 |
| 244 | 27/02/2015 | 15:39:30 |   0 |   19 |
| 245 | 27/02/2015 | 15:39:30 |   1 |   16 |
| 246 | 27/02/2015 | 15:39:30 |   2 |   16 |
| 247 | 27/02/2015 | 15:40:30 |   0 |   19 |
| 248 | 27/02/2015 | 15:40:30 |   1 |   16 |
| 249 | 27/02/2015 | 15:40:30 |   2 |   16 |
| 250 | 27/02/2015 | 15:41:30 |   0 |   19 |
| 251 | 27/02/2015 | 15:41:30 |   1 |   16 |
| 252 | 27/02/2015 | 15:41:30 |   2 |   16 |
| 253 | 27/02/2015 | 15:42:30 |   0 |   18 |
| 254 | 27/02/2015 | 15:42:30 |   1 |   18 |
| 255 | 27/02/2015 | 15:42:30 |   2 |   18 |
+-----+------------+----------+-----+------+
```

Figure 9-27   Portion of `sensorTemp` records created by the SensorDatabase program.

You can now start the MySQL program and look at the records generated by entering the following SQL command at the mysql prompt:

SELECT * FROM sensorTemp;

Figure 9-27 shows a portion of the `sensorTemp` table results after I entered the above command. You can also view the records by running the following program named ViewRecords.py. It is also available on the book's companion website.

```
#!/usr/bin/python
import MySQLdb
#this creates the basic connection object
db = MySQLdb.connect(host='localhost', user='TempUser1',
passwd='Px158qqr', db='HomeTempSystem')
```

```
#a cursor object is required to execute SQL commands on the database
cur = db.cursor()
#this SQL command retrieves all the records from the sensorTemp table
cur.execute('SELECT * FROM sensorTemp')
#display all the data record after record
for row in cur.fetchall():
    print row[0], "   ", row[1], "   ", row[2], "   ", row[3], "   ",
row[4]
```

Execute the ViewRecords.py program by entering the following at the command line:

```
python ViewRecords.py
```

Figure 9-28 shows the resulting display after the program is run. Only a portion of the records that were generated by the SensorDatabase.py program is displayed.

```
214    27/02/2015    15:29:29    0    19
215    27/02/2015    15:29:29    1    16
216    27/02/2015    15:29:29    2    16
217    27/02/2015    15:30:29    0    19
218    27/02/2015    15:30:29    1    16
219    27/02/2015    15:30:29    2    16
220    27/02/2015    15:31:29    0    18
221    27/02/2015    15:31:29    1    16
222    27/02/2015    15:31:29    2    17
223    27/02/2015    15:32:29    0    19
224    27/02/2015    15:32:29    1    16
225    27/02/2015    15:32:29    2    17
226    27/02/2015    15:33:29    0    20
227    27/02/2015    15:33:29    1    16
228    27/02/2015    15:33:29    2    17
229    27/02/2015    15:34:29    0    18
230    27/02/2015    15:34:29    1    16
231    27/02/2015    15:34:29    2    17
232    27/02/2015    15:35:29    0    18
233    27/02/2015    15:35:29    1    16
234    27/02/2015    15:35:29    2    17
235    27/02/2015    15:36:30    0    19
236    27/02/2015    15:36:30    1    15
237    27/02/2015    15:36:30    2    16
238    27/02/2015    15:37:30    0    18
239    27/02/2015    15:37:30    1    15
240    27/02/2015    15:37:30    2    17
241    27/02/2015    15:38:30    0    19
242    27/02/2015    15:38:30    1    16
243    27/02/2015    15:38:30    2    17
244    27/02/2015    15:39:30    0    19
245    27/02/2015    15:39:30    1    16
246    27/02/2015    15:39:30    2    16
247    27/02/2015    15:40:30    0    19
248    27/02/2015    15:40:30    1    16
249    27/02/2015    15:40:30    2    16
250    27/02/2015    15:41:30    0    19
251    27/02/2015    15:41:30    1    16
252    27/02/2015    15:41:30    2    16
253    27/02/2015    15:42:30    0    18
254    27/02/2015    15:42:30    1    18
255    27/02/2015    15:42:30    2    18
root@ubilinux:~#
```

Figure 9-28   *SensorDatabase-generated records displayed by the ViewRecords program.*

I next modified the ViewRecords program to display only the records from channel 0. This modification required inserting a conditional phrase in the SQL SELECT statement:

```
cur.execute('SELECT * FROM sensorTemp WHERE tchan = 0')
```

I renamed ViewRecords.py to Chan0ViewRecords.py. I did not provide a program listing, as it required only that one slight change to the SELECT statement. To run it, simply enter:

```
python Chan0ViewRecords.py
```

Figure 9-29 shows the output from the program in which only the channel 0 records are displayed.

```
130    27/02/2015    15:01:24    0    20
133    27/02/2015    15:02:24    0    20
136    27/02/2015    15:03:25    0    19
139    27/02/2015    15:04:25    0    19
142    27/02/2015    15:05:25    0    19
145    27/02/2015    15:06:25    0    19
148    27/02/2015    15:07:25    0    19
151    27/02/2015    15:08:25    0    19
154    27/02/2015    15:09:26    0    19
157    27/02/2015    15:10:26    0    19
160    27/02/2015    15:11:26    0    19
163    27/02/2015    15:12:26    0    19
166    27/02/2015    15:13:26    0    19
169    27/02/2015    15:14:26    0    19
172    27/02/2015    15:15:26    0    19
175    27/02/2015    15:16:27    0    20
178    27/02/2015    15:17:27    0    18
181    27/02/2015    15:18:27    0    19
184    27/02/2015    15:19:27    0    19
187    27/02/2015    15:20:27    0    19
190    27/02/2015    15:21:27    0    18
193    27/02/2015    15:22:27    0    19
196    27/02/2015    15:23:28    0    19
199    27/02/2015    15:24:28    0    18
202    27/02/2015    15:25:28    0    19
205    27/02/2015    15:26:28    0    19
208    27/02/2015    15:27:28    0    19
211    27/02/2015    15:28:28    0    18
214    27/02/2015    15:29:29    0    19
217    27/02/2015    15:30:29    0    19
220    27/02/2015    15:31:29    0    18
223    27/02/2015    15:32:29    0    19
226    27/02/2015    15:33:29    0    20
229    27/02/2015    15:34:29    0    18
232    27/02/2015    15:35:29    0    18
235    27/02/2015    15:36:30    0    19
238    27/02/2015    15:37:30    0    18
241    27/02/2015    15:38:30    0    19
244    27/02/2015    15:39:30    0    19
247    27/02/2015    15:40:30    0    19
250    27/02/2015    15:41:30    0    19
253    27/02/2015    15:42:30    0    18
root@ubilinux:~#
```

Figure 9-29 *SensorDatabase*-generated records displayed by the *Chan0ViewRecords* program.

```
0    Case    240    27/02/2015    15:37:30
1    Workshop    240    27/02/2015    15:37:30
2    Outside    240    27/02/2015    15:37:30
0    Case    241    27/02/2015    15:38:30
1    Workshop    241    27/02/2015    15:38:30
2    Outside    241    27/02/2015    15:38:30
0    Case    242    27/02/2015    15:38:30
1    Workshop    242    27/02/2015    15:38:30
2    Outside    242    27/02/2015    15:38:30
0    Case    243    27/02/2015    15:38:30
1    Workshop    243    27/02/2015    15:38:30
2    Outside    243    27/02/2015    15:38:30
0    Case    244    27/02/2015    15:39:30
1    Workshop    244    27/02/2015    15:39:30
2    Outside    244    27/02/2015    15:39:30
0    Case    245    27/02/2015    15:39:30
1    Workshop    245    27/02/2015    15:39:30
2    Outside    245    27/02/2015    15:39:30
0    Case    246    27/02/2015    15:39:30
1    Workshop    246    27/02/2015    15:39:30
2    Outside    246    27/02/2015    15:39:30
0    Case    247    27/02/2015    15:40:30
1    Workshop    247    27/02/2015    15:40:30
2    Outside    247    27/02/2015    15:40:30
0    Case    248    27/02/2015    15:40:30
1    Workshop    248    27/02/2015    15:40:30
2    Outside    248    27/02/2015    15:40:30
0    Case    249    27/02/2015    15:40:30
1    Workshop    249    27/02/2015    15:40:30
2    Outside    249    27/02/2015    15:40:30
0    Case    250    27/02/2015    15:41:30
1    Workshop    250    27/02/2015    15:41:30
2    Outside    250    27/02/2015    15:41:30
0    Case    251    27/02/2015    15:41:30
1    Workshop    251    27/02/2015    15:41:30
2    Outside    251    27/02/2015    15:41:30
0    Case    252    27/02/2015    15:41:30
1    Workshop    252    27/02/2015    15:41:30
2    Outside    252    27/02/2015    15:41:30
0    Case    253    27/02/2015    15:42:30
1    Workshop    253    27/02/2015    15:42:30
2    Outside    253    27/02/2015    15:42:30
0    Case    254    27/02/2015    15:42:30
1    Workshop    254    27/02/2015    15:42:30
2    Outside    254    27/02/2015    15:42:30
0    Case    255    27/02/2015    15:42:30
1    Workshop    255    27/02/2015    15:42:30
2    Outside    255    27/02/2015    15:42:30
root@ubilinux:~#
```

Figure 9-30 *SensorDatabase*-generated records displayed by the ViewRecords program with the *channelLocation* table included.

To view the sensor location text descriptions along with the temperature data, change the SELECT statement in the ViewRecords program to the following:

```
cur.execute('SELECT * FROM  channelLocation,
sensorTemp')
```

Figure 9-30 shows a portion of the output from the program in which the location description is shown along with the temperature data. Notice that I re-ordered the placement of the two table fields in the SQL query. I found that the query did not work properly when sensorTemp was first in the order. The reason is a bit complex, having to do with the way MySQL handles the keys between joined tables.

The previous demonstration finishes the discussion of creating database records from a sensor-based acquisition program. The next phase involves showing how to access and display selected records using a web browser.

Database Access Using a Web Browser

This is probably one of the easier portions of the project, as it concerns the well-documented process of creating a dynamic website (HTML) that uses a web server language (PHP) to supply data from a relational database (MySQL) to a client (remote browser) upon demand. I am not including a discussion of *record locking* for now, which is a solution to the problem of two applications trying to access the same database record at the same time. I will just make the reasonable assumption that the database records are all accessible whenever the web server application needs them.

I will initially present a direct approach to creating a website; however, there are many books and online tutorials available on different approaches to creating a dynamic website. This website will not be fancy or flashy but simply serve up the desired temperature records in a tabular format.

The following is some straightforward PHP5 code that will display all the records in the sensorTemp table, which is part of the HomeTempSystem database. Notice that I logged in as TempUser1, which was added for security reasons after the creation of the database. To run it as shown, you will need to store this script as TempSensorTest.php in the /var/www directory.

```php
<?php
$username = "TempUser1";
$password = "Px158qqr";
$hostname = "localhost";
$database = "HomeTempSystem";

//connection to localhost
$con = mysqli_connect($hostname, $username, $password, $database);

//execute the SQL query and return records
$result = mysqli_query($con, "SELECT * FROM sensorTemp");

//fetch the data from the result recordset
while ($row = mysqli_fetch_array($result))
  {
  echo $row['id']." ".$row['tdate']." ".$row['ttime'].
" ".$row['tchan']." ".$row['ttemp'];
```

```
214 27/02/2015 15:29:29 0 19
215 27/02/2015 15:29:29 1 16
216 27/02/2015 15:29:29 2 16
217 27/02/2015 15:30:29 0 19
218 27/02/2015 15:30:29 1 16
219 27/02/2015 15:30:29 2 16
220 27/02/2015 15:31:29 0 18
221 27/02/2015 15:31:29 1 16
222 27/02/2015 15:31:29 2 17
223 27/02/2015 15:32:29 0 19
224 27/02/2015 15:32:29 1 16
225 27/02/2015 15:32:29 2 17
226 27/02/2015 15:33:29 0 20
227 27/02/2015 15:33:29 1 16
228 27/02/2015 15:33:29 2 17
229 27/02/2015 15:34:29 0 18
230 27/02/2015 15:34:29 1 16
231 27/02/2015 15:34:29 2 17
232 27/02/2015 15:35:29 0 18
233 27/02/2015 15:35:29 1 16
234 27/02/2015 15:35:29 2 17
235 27/02/2015 15:36:30 0 19
236 27/02/2015 15:36:30 1 15
237 27/02/2015 15:36:30 2 16
238 27/02/2015 15:37:30 0 18
239 27/02/2015 15:37:30 1 15
240 27/02/2015 15:37:30 2 17
241 27/02/2015 15:38:30 0 19
242 27/02/2015 15:38:30 1 16
243 27/02/2015 15:38:30 2 17
244 27/02/2015 15:39:30 0 19
245 27/02/2015 15:39:30 1 16
246 27/02/2015 15:39:30 2 16
247 27/02/2015 15:40:30 0 19
248 27/02/2015 15:40:30 1 16
249 27/02/2015 15:40:30 2 16
250 27/02/2015 15:41:30 0 19
251 27/02/2015 15:41:30 1 16
252 27/02/2015 15:41:30 2 16
253 27/02/2015 15:42:30 0 18
254 27/02/2015 15:42:30 1 18
255 27/02/2015 15:42:30 2 18
```

Figure 9-31   *Portion of the* `sensorTemp` *records displayed on a separate networked computer.*

```
echo "<br>";
}

//close the connection
mysql_close($con);
?>
```

You should try to access the records using a separate computer attached to your home network. I used my Macbook Pro and put in the URL `http://192.168.0.6/SensorTempTest.php`. Figure 9-31 shows the result of this action.

Note, your local IP address for the dev board will likely be different from my address. Just substitute whatever your address is. Remember, you can always find it by entering:

```
ifconfig
```

Just look for the IP address next to the wlan0 entry for wireless connection.

Narrowing the Database Reports

It is not hard to imagine that the database size will grow rapidly as you accumulate more measurements over time. It would be a waste of time and bandwidth to have to sift through all the database records to find the specific data that you needed to examine. I will now show you how to apply constraints to the database search by using the "WHERE" phrase that I introduced earlier. The major complication is that you will be using a web browser to access the database and, therefore, will not be able to insert the WHERE phrase in the SQL query directly, as was done in the earlier demonstration. Fortunately, this situation is well covered in the HTML and PHP languages in which a form will be created asking you for specific information to be sent to the web server for action.

In this example, I am asking only for records of a specific channel number to be displayed; however, the concepts can be readily extended to all the other database fields. There are two primary means by which data get sent by a user to a web server. These are the GET and POST methods. Each has their advantages and disadvantages, but I have found that most developers prefer using the POST method so that is what I will implement. The code for a simple HTML form requesting a channel number is shown below:

```
<html>
<body>
<form action = http://192.168.0.6/TempSensorTestChan.php
 method = "POST">
Channel number: <input type = "text" name = "chan_no"><br>
<input type = "submit">
</form>
</body>
</html>
```

I named this code ChannelSelector.html and saved it on my laptop in the Documents folder. This program is really more of a script that the remote client browser will use to send and receive data from the Edison's web server. You should also notice that I hardcoded the server's IP address into the form along with a reference to a slightly modified version of the three-channel TempSensorTest program,

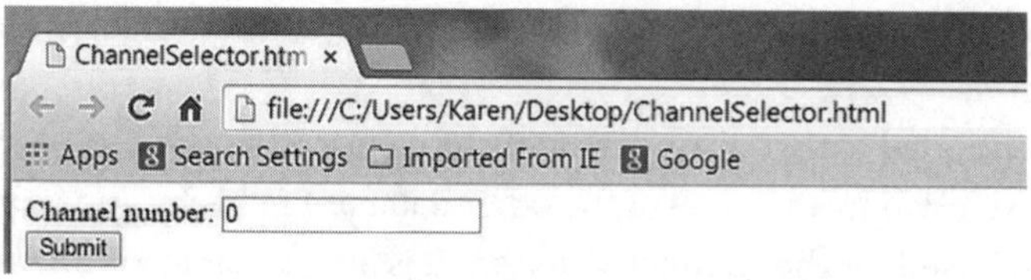

Figure 9-32 *ChannelSelector* form.

which I discuss below. The key point regarding this HTML script is that it has a variable named "chan_no" that stores the channel number and this variable is made available on the server side by the POST method. Figure 9-32 shows the form on a client-side browser.

Nothing will happen regarding database access until the submit button is clicked. I will show the results of clicking on the button after I finish with the server program.

The server-side program is named TempSensorTestChan.py to which the suffix "Chan" was added to indicate that the program constrains its output to the user-selected channel. The code shown below is identical to the TempSensorTest program with some modifications to accommodate the user-selected channel.

```php
<?php
$username = "TempUser1";
$password = "Px158qqr";
$hostname = "localhost";
$database = 'HomeTempSystem';

//connection to localhost
$con = mysqli_connect($hostname, $username, $password, $database);

//channel request
$channel = $_POST['chan_no'];

//execute the SQL query and return selected records
$result = mysqli_query($con, "SELECT * FROM sensorTemp
WHERE tchan = $channel");

//fetch the data from the result recordset
while ($row = mysqli_fetch_array($result))
  {
  echo $row['id']." ".$row['tdate']." ".$row['ttime'].
" ".$row['tchan']." ".$row['ttemp'];
  echo "<br>";
  }

//close the connection
mysql_close($con);
?>
```

Figure 9-33 shows the browser display after channel 0 was selected via the input form.

As I mentioned earlier, the constraints to be placed in the WHERE phrase can easily be extended to any or all of the other database fields by applying the same techniques I used for the channel selection. It is also possible to construct a complete query statement on the client side and pass that over to the server, but that's best left to a more advanced study of web-based database retrieval. I found this technique quite suitable for these types of embedded applications.

This last section concludes my discussion of databases and web servers, which can be implemented using the Edison running the Debian distribution.

Figure 9-33   *Portion of the channel 0* sensorTemp *records displayed on a separate networked computer.*

Summary

I covered a lot of material in this lengthy chapter; however, I believe it will greatly benefit you to attempt to duplicate some or all of the projects presented.

Setting up a LAMP project is a very fundamental task that is often encountered in real-world situations. Getting to know how to use the Apache web server with PHP is a good skill that will carry over to many different areas, not just using an embedded processor.

I introduced you to some very basic and important database fundamentals by using the MySQL database for the home temperature system. This should really whet your appetite to learn more about how to use databases, as they are the backbone for any technical or business system.

Learning how to programmatically insert data into a database is another good experience builder. Knowing how to use a database with an embedded system is important, as I am fairly confident that is the method most embedded systems use to save any acquired data.

10
Wearables

In this chapter, I will discuss a relatively new technology area loosely termed as *wearables* that concerns portable electronic systems that can be worn as part of clothing or attached to the body with straps or to clothing with clip-ons. The Edison module is an ideal form factor to be part of a wearable system because of its small size, low power consumption, wireless communication, and stackable physical architecture. I will also be using the stackable modules that are part of the Sparkfun kit designed for the Edison.

Sparkfun Console Module

Figure 10-1 shows Sparkfun's base module, which provides all the connectivity required for the Edison module. It has two micro-USB connectors, one that provides power and the other USB connectivity to the Edison that is mounted via a 70-pin Hirose connector on the back side of the base module. I will also be using an Edison with the Debian distribution already loaded into it for the chapter project. It is important that the Edison's WiFi be configured properly, as that is a key project component.

Stackable Architecture

I use the phrase *stackable architecture* to refer to how the physical Sparkfun modules connect to each other. The Edison is mounted on a base module for this chapter's project and other modules are connected, or stacked, via two 70-pin Hirose connectors, one male and the other female, that are mounted on each side of a module. There is a 3.5-mm spacing between modules when they are stacked,

Figure 10-1 *Sparkfun Base module.*

which should provide sufficient clearance for all the surface-mounted components. Sparkfun also has a very inexpensive hardware kit, which is shown in Figure 10-2.

I strongly recommend using the hardware kit to firmly secure all the stacked modules; otherwise, they would be held in-place only by the pin contact pressure from the Hirose connectors. I am positive that leaving the modules attached by only the Hirose connectors would eventually lead to problems, especially if the stack is put into motion, which is the case for this chapter's project.

Figure 10-2 *Sparkfun Hardware kit.*

Chapter Project

I decided on a relatively simple project to demonstrate both the wearability and utility of the Edison. The project is a movement monitor that would alert caregivers if an infirm or disabled patient gets out of bed and starts moving around without any assistance. A patient moving without assistance in the dark could easily fall, which, statistically, is a major reason for elderly morbidity. A patient getting out of bed at night is a high risk situation, which normally would require caregivers to be on alert. Staying awake all night is a very poor alternative to using an automated system, which would awaken the caregiver when the patient gets up and attempts to move about.

This project uses stacked Edison modules consisting of:

- Base module with Edison attached
- Battery module
- 9DOF sensor module

Figure 10-3 is the project block diagram showing the three modules. Of course, all the modules are interconnected using the Hirose connectors.

I will next discuss the battery module, as I have already discussed the base module.

Battery Module

This module provides all the power necessary for the other two modules. Its primary power source is a 3.7-V lithium polymer (LiPo) battery, which may be charged using a micro USB connector on the module. Figure 10-4 shows this module.

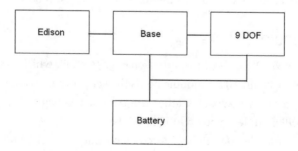

Figure 10-3 *Block diagram.*

Figure 10-4 *Battery module.*

There is a significant issue with the battery module's charge capacity. The LiPo battery on the module is rated for 400 milliampere hours (mAh). This means it will provide 400 ma for one hour before becoming exhausted. This specification is related to battery design and construction and cannot be modified. The only recourse is to minimize power consumption from all the other stacked modules. The base module with the Edison mounted on it is by far the greatest power consumer. Running minimal code and essentially idling, it takes about 200 ma. When the WiFi is operating, it can consume as much as 800 ma. This means that the project software must carefully control the WiFi, turning it on only when it is needed. The single battery life can then be extended to approximately two hours, which, while not optimal, is far better than 30 minutes with the WiFi constantly operating.

Another option for extending operational time is to use a larger-capacity battery. This will make the wearable device quite a bit larger and bulkier but should significantly extend the operating time for all the stacked modules.

9DOF Module

The descriptor 9DOF is short for *nine degrees of freedom*, which is the module's measurement capability. This module holds the sensor that, in combination with a Python program, will sense whether the patient has begun to get up from a bed. The 9DOF module is shown in Figure 10-5.

The 9DOF module uses the ST Microelectronics model LSM9DS0 inertial measurement unit (IMU) to create a full range-of-motion sensor. This chip

Figure 10-5 *9DOF module.*

contains a 3-axis accelerometer, a 3-axis gyroscope, and a 3-axis magnetometer. The chip communicates to the Edison by using the I2C bus. Table 10-1 shows the sensor ranges for all three measurement types.

Note that little "g" in the table is for a *unit of gravitation*, while capital "G" stands for *gauss*, which is the SI unit for magnetic flux density. For your information, the Earth's magnetic field at its surface ranges from 0.31 to 0.58 gauss when measured parallel to the lines of force. It is entirely possible to determine if a patient is attempting to get out of bed by sampling all three measurement types, each of which has three orthogonal axes. *Orthogonal* means the axes of the measurements are oriented 90° to each other.

One major requirement for this project was to minimize the generation of false alarms, i.e., patient motion that might resemble getting-out-of-bed behavior but is really only random twitching or normal sleep motions that trigger the threshold levels used in the program, all of which were empirically derived. The threshold level that I finally decided to use was the result of observing the data generated by the Fusion test software, which I describe later. Ultimately, it was simply a trial-and-error approach to set the level empirically.

Measurement Type	Ranges
Acceleration	± 2g, ± 4g, ± 6g, ± 8g, ± 16g
Gyroscopic	± 245°/s, ± 500°/s, ± 2000°/s
Magnetic	± 12G

Table 10-1 *LSM9DS0 Measurement Ranges*

The Project Software

I started the software design with a conceptual flowchart that is shown in Figure 10-6. I have used this approach for many projects, and it seems to work for me. Others may have different ways to conceptualize a project, which is fine as long as they produce the desired results.

The flowchart blocks encapsulate the project's requirements in a reasonable flow process. I have outlined below the requirements in each block. Note that the event I refer to in the block discussions is the patient getting-out-of-bed situation.

- *Initialization*—In this block, I need to ensure that the sensor is functioning as desired and that the WiFi is also set up properly.
- *Acquire data*—This block samples the 9DOF sensor for all of the generated signals. The big question is to determine how often to sample. Sampling constantly will be a bigger drain on the battery than periodic sampling; however, you don't want to miss the event. I think that sampling every second might be a good initial guess. Proceed with acquiring new sensor data if the parameters are not exceeded
- *Exceed params*—This block does an immediate check of all of the sensor values to see if a selected measurement exceeds the preset threshold, which would indicate that the event has occurred.

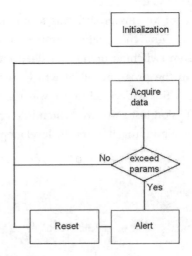

Figure 10-6 *Conceptual flowchart.*

- *Alert*—Transmit a message that an event has been detected, using the WiFi. I might do this several times just to ensure that the message is not lost.
- *Reset*—Reset any parameters that were set while the previous blocks were executed. Proceed with acquiring new sensor data

I also used the open-source RTIMUlib library created by richards-tech.com. This library contained all the Python code I needed to both configure and read the sensor data generated by the 9DOF module. This library may be downloaded from GitHub by searching for richards-tech/RTIMULib for Linux. There is also a version for the Arduino, which, I am pretty sure, will not work with the Debian distribution. The library also contains C/C++ code for readers interested in those languages.

I transferred the complete Python subdirectory to the Edison while it was still mounted on the dev board. Once you have the Python directory transferred, you will need to build and install the library. Thankfully, this is easily handled by a script in the directory named setup.py. Just enter the following commands at the root level:

```
python setup.py build
python setup.py install
```

That should be all you need to do to have the library ready for your use. I checked the installation by going into the tests subdirectory and running a script named Fusion.py. Simply enter:

```
python Fusion.py
```

You should see a tabular list of three parameters, r, p, and y, scrolling rapidly on the terminal screen, all with zero values, as there is no 9DOF sensor attached to the dev board yet. The sensor is a stackable form factor module and must be installed in the stack before any data can be generated. There is one more key application that must be installed before the stack can be fully functional. I discuss this application next.

sudo

sudo is an application that permits a non-root user to execute as if the user were at the root level for one command at a time. Previously, in this book, I repeatedly mentioned that you should be at the root level to do a specific command. That required you to enter the su command along with the root password. Now, by using sudo, you can remain at the previous user level and simply prepend the desired command with sudo to carry out the desired superuser operation. For example, simply enter sudo nano <filename> to do an edit in lieu of switching to the root level where you entered only nano <filename>.

You will first need to switch to the root level in order to download and install the sudo application, which is done by entering:

```
apt-get install sudo
```

After sudo is installed you will next need to alter the sudo configuration file, which is named sudoers and is located in the etc directory. The sudoers file is normally edited by using a special editor named visudo; however, I will show you how to use the nano editor instead, as that is much friendlier than using visudo, which is based on the classic vi editor. Enter the following to change the sudoer editor to nano:

```
export VISUAL=nano visudo
```

Now you are ready to alter the sudoers file by first going to the etc directory and entering:

```
nano sudoers
```

Now enter the following after the existing line containing root ALL= (ALL:ALL) ALL

```
edison ALL=(ALL) NOPASSWD:ALL
```

Save the file and exit the nano editor, and you should be all set to use the sudo command.

The Project Stack

Figure 10-7 shows front and back views of all of the project modules connected to the stack. The modules are also physically connected using the hardware kit components to increase their connection's strength and rigidity.

There is only one way to stack the modules due to the clearances involved with each module. I would also strongly recommend you use the module stacking hardware, as that relieves most of the stresses on the Hirose connectors. You mount the Edison on the Base Block and then mount the 9DOF Block to the back of the Base. Next mount the Battery Block to the back of the 9DOF Block. The

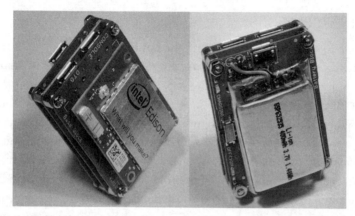

Figure 10-7   *Project stack.*

assembly is fairly easy, but be careful with the tiny hardware spacers, screws, and nuts, as they are easy to lose.

I also used two spacers with screws and nuts to secure the Edison module to the Base Block to prevent the Edison from rocking in its socket, which eventually might lead to connection problems.

Initial Project Stack Test

Just connect one micro USB cable from the connector on the Base Board labeled console to your laptop running a terminal program. Remember you will have to determine the USB serial name as you have done before, using the following command:

```
ls /dev/tty* | more
```

I added the more command as a Linux pipe because there are many tty devices and the one you are looking for is at the top of the list and will scroll by before you can read it. Just hold down the enter (return) key to scroll through the rest of the device listing after you have written down the USB device name. In my case, I entered the following to get to the initial login screen:

```
screen /dev/tty.usbserial-DA017Q4Y 115200 -L
```

Your login will be different because each Base Board has its own unique identification. The stack behaves exactly like it did when you were connected to the dev board. My next step was to halt the computer and disconnect the USB, since it was time to check out the battery operation.

Battery Operations

Being able to operate the stack using the battery is key to having a wearable application. Of course, you cannot use a direct USB connection anymore so it was necessary to login remotely to the stack, using a Windows desktop computer and the putty application, to continue the software development. You might need to go back to Chapter 6 to refresh your memory about using putty. Obviously, you must power on the stack, which is easily done by using the micro slide switch found on the Battery Block. Wait about 30 seconds for the Edison to establish a WiFi connection before trying to connect with the putty application.

I have found it much easier to develop and modify software using a remote ssh login as compared to using a direct USB connection. This is because the nano editor tends to be a bit "cranky" when used with a direct USB connection, but behaves quite properly when used with an ssh connection. I am not quite sure why this happens, but I do appreciate using nano with the ssh login.

The first test I tried after successfully connecting was to run the Fusion.py script again. This time it showed actual data, since the 9DOF sensor was now part of the stack. I ran the script by entering:

```
sudo python /root/RTIMULib-master/Linux/python/tests/
Fusion.py
```

Figure 10-8 shows the results of this command.

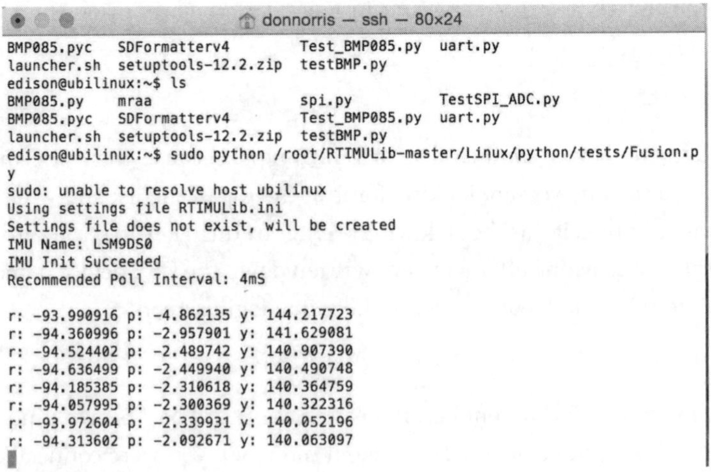

Figure 10-8 *Fusion.py program running.*

It is now time to discuss how to transmit or publish the data generated by this Edison stack so that interested parties can access it and be alerted when the patient is on the move. I will be using the MQTT messaging protocol, which is part of the Paho project.

Paho and Eclipse.org

Paho is an open-source project sponsored by the Eclipse.org foundation. This project is dedicated to providing scalable client implementations for both open and standard messaging protocols. The Paho project is designed to provide an exciting infrastructure in support of the new Machine to Machine (M2M) and Internet of Things (IoT) applications. The home website is:

http://www.eclipse.org/paho/

At the heart of the Paho project is a lightweight publish/subscribe message protocol named MQTT, which I describe in the next section.

MQTT

MQTT is the current name for this protocol, although it was originally named Message Queuing Telemetry Transport. I guess the project managers felt that was a mouthful, or it could be the fact that there are no actual queues used in the protocol. In any case, it is now simply called MQTT.

MQTT is over 11 years old, having originally been created by the IBM Pervasive software group in conjunction with Arcom, which is now called Eurotech. IBM still supports MQTT with the current 3.1 version specifications available from the IBM developerWorks website at:

http://www.ibm.com/developerworks/webservices/library/ws-mqtt/index.html

MQTT is technically known as a middleware application, as can be seen in the Figure 10-9 block diagram for this project. It is important to realize that both publishers and subscribers are treated as client applications in this configuration type.

There is a block named MQTT Broker located between the Edison publisher-client and the subscriber-client blocks. The clients may be a PC, Mac, Android, or iOS device, as shown in Figure 10-9. The broker shown may be thought of as a message dispatcher ensuring that the MQTT messages are properly sent from

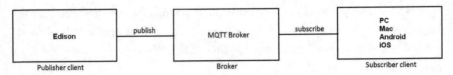

Figure 10-9   *Project block diagram.*

the client publishers to the correct client subscribers. In that way, subscriber clients do not have to constantly monitor all the network traffic looking for messages that are addressed to them. The broker takes over that function, and it also serves as an acknowledgment intermediary, which I explain in the section concerning quality of service.

Table 10-2 details some of the MQTT salient features that make it so popular as a messaging protocol. These features make MQTT very popular for M2M applications, including weather monitoring, stock tickers, smart power grid meters, and even Facebook messaging. It is also a very popular way for cellular services to implement message alerts.

Quality of Service (QoS)

Quality of Service (QoS) refers to the level of assurance that MQTT provides regarding message delivery. There are three QoS levels:

1. *Level 0*—This is also known as "fire and forget." At this level, the publisher sends off messages, and there is no attempt to acknowledge their reception by the broker on behalf of the publisher. It is obviously the quickest message delivery method, but it is also the least reliable.

2. *Level 1*—This is also known as "at least one." Here, messages are sent and resent until the broker receives one acknowledgement from the subscriber. It does provide some assurance that the message did get

Feature	Description
API	Simple, only five methods required
Packets	Compact binary packets. Capable of up to 250MB payload
Headers	No compressed headers needed
Verbose	Minimal text, much less than HTML

Table 10-2   *MQTT Features*

through to its intended recipient. Level 1 is typically set as the default QoS for a MQTT messaging system.

3. *Level 2*—This is also known as "exactly one." At this level, messages undergo a two-stage process in which there is a definitive acknowledgement between the broker and subscriber, ensuring that one, and only one, message copy was delivered. This QoS level is the slowest among the three levels due to the additional processing overhead required to establish a high reliability level.

Wills

No, this section has nothing to do with legal probate but instead focuses on what happens when a client abnormally loses its connection with the broker. A *will* is both a set of instructions and a prescribed message that is stored by the broker and will be acted upon only if the connection between the broker and a client is unexpectedly broken. Basically, it is a dialog that states, "If you (the broker) cannot connect to me, and I (the client) haven't cleanly disconnected, then carry out the preset instructions and also send out the stored message on my behalf." The "will" concept is implemented in Python by a `setWill` method and in Java by an object of the `MqttConnectionsOptions` class.

Using wills in a MQTT protocol improves both system robustness and reliability and ensures that either messages will be delivered or an error message describing what went wrong will be created and distributed.

Reconnecting

Connections will be broken, and MQTT has the inherent ability to reconnect using two system elements. The first is a logical flag known as the *clean session flag*, which is set for every fresh or new connection. The clean session flag informs the client and broker that they must start the messaging process from the beginning because it represents a new connection. However, if the clean session flag becomes false or low, a second element comes into play. This is called the client ID, and as we shall see when the test code is discussed, it plays a key part in establishing an original connection.

For now, let's assume the clean session flag had already been set to some `String` value when the connection broke. Now, assume the connection is restored, as might happen when a client briefly loses power. MQTT will attempt to restore the connection to the same precise state as before the disconnection because it recognizes it still has stored in its record structure the same client ID `String` that existed for this particular connection when it first became disconnected. Note, that various MQTT libraries, whether they are Python or Java, have different implementations for storing the client IDs and messages so that the connections can be recovered without any message loss.

It is now time to go back to the patient-movement monitoring program and incorporate MQTT to distribute single-value data points between one publisher and one client.

Edison MQTT Publisher Client

You first need to load the appropriate MQTT client implementation library before you can add the messaging features to the application. Please follow these steps to load the Python library that will be used in this project.

1. The Debian distribution must first be updated to ensure that all dependencies will be located in the appropriate repositories. Enter the following:
   ```
   sudo apt-get update
   ```

2. If you do not have git already installed, you will first need to enter this command:
   ```
   sudo apt-get install git
   ```

3. Download the source code from GitHub:
   ```
   sudo git clone
   git://git.eclipse.org/gitroot/paho/org.eclipse
   .paho.mqtt.python.git
   ```
   NOTE   If you have difficulty in doing a direct `git` `clone` *you may also go to:* `http://git.eclipse.org/c/paho/org.eclipse` `.paho.mqtt.python.git/` *and download one of the following compressed files:*
   ```
   org.eclipse.paho.mqtt.python-1.0.zip
   org.eclipse.paho.mqtt.python-1.0.tar.gz
   org.eclipse.paho.mqtt.python-1.0.tar.bz2
   ```
   Use the extraction application that matches the compressed file extension you downloaded, i.e., winzip or 7zip for a zip file. There should be

the same source directory created after extraction as there was for the clone operation.

4. Change into the source directory:
   `cd org.eclipse.paho.mqtt.python/`

5. Compile the source code using a build script already available in the directory:
   `sudo make`

6. Install all the compiled files:
   `sudo make install`

The Python MQTT client should now be ready to be added to the Fusion.py program. However, I will first cover some basic concepts that you should have clear in your mind before going on to the complete application.

Edison publisher client must establish a logical connection to the broker before any messages can be passed. This is done with the following statements:

`import paho.mqtt.client as mqtt`—sets up a client reference named `mqtt`

`mqttc = mqtt.Client()`—instantiates an MQTT client object named `mqttc`

`mqttc.connect("m2m.eclipse.org", 1883, 60)`—Goes out to the Internet and connects with an MQTT broker at the website "m2m `.eclipse.org`" on port `1883`. The `60` refers to a 60-second ping that is a *keep-alive*, meaning it is sent when no other activity is happening on the connection.

`mqttc.loop_start()`—A separate execution thread is started that handles incoming messages from the broker.

The following statement contains references to what are known as *topics* and *subtopics*.

`mqttc.publish("edison123/imu/p","%.2f" % pose);`

"edison123" in the above statement refers to a `root` topic created on the broker that also contain the subtopics, `imu` and `p`. The `pose` data, representing the `p-coordinate` data, is stored in the `p` subtopic that is also a subtopic of the `imu` subtopic. I will shortly demonstrate how to retrieve this real-time data from the broker. But first, you should enter the following modified Fusion.py program that I renamed mqttFusion.py to reflect the new messaging capabilities:

```python
import sys, getopt
sys.path.append('.')
import RTIMU
import os.path
import time
import math
import paho.mqtt.client as mqtt

SETTINGS_FILE = "RTIMULib"

print("Using settings file " + SETTINGS_FILE + ".ini")
if not os.path.exists(SETTINGS_FILE + ".ini"):
  print("Settings file does not exist, will be created")

s = RTIMU.Settings(SETTINGS_FILE)
imu = RTIMU.RTIMU(s)

mqttc = mqtt.Client()
mqttc.connect("m2m.eclipse.org", 1883, 60)
mqttc.loop_start()

print("IMU Name: " + imu.IMUName())

if (not imu.IMUInit()):
    print("IMU Init Failed");
    sys.exit(1)
else:
    print("IMU Init Succeeded");

poll_interval = imu.IMUGetPollInterval()
print("Recommended Poll Interval: %dmS\n" % poll_interval)

while True:
  if imu.IMURead():
    # x, y, z = imu.getFusionData()
    # print("%f %f %f " % (x,y,z) " mqtt")
    data = imu.getIMUData()
    fusionPose = data["fusionPose"]
    print("r: %f p: %f y: %f mqtt" % (math.degrees(fusionPose[0]),
        math.degrees(fusionPose[1]), math.degrees(fusionPose[2])))
    time.sleep(poll_interval*300.0/1000.0)

  pose = math.degrees(fusionPose[1])
    if(math.fabs(pose) > 15.0):
        mqttc.publish("edison123/imu/p", "%.2f" % pose);
```

The program is run by entering:

```
sudo python /root/RTIMULib-master/Linux/python/tests/
mqttFusion.py
```

You should see exactly the same terminal display that was shown when the previous Fusion.py program was run in the earlier test except that I added the word mqtt to the end of each display line to help me distinguish between the two program outputs. Figure 10-10 is a terminal screenshot for this program.

Any p data exceeding 15.0° is also being sent to the broker located at m2m.eclipse.org and listening on port 1883. This 15.0° value is simply my best estimate at indicating patient movement based on the fact that the sensor would be in a nearly vertical position when the patient stood, assuming that the stack was placed with its long axis parallel to the patient's leg. The exact positioning is not critical because the nominal sensor level will be near 0 while the patient is lying down. The 15.0° can easily be changed if it is found to generate too many false alarms

The next section concerns how to auto start the mqttFusion.py script upon initial power on, since that will be the way it will normally be operated.

```
r: 142.219688 p: -7.498789 y: -61.136491 mqtt
r: 142.287334 p: -7.728905 y: -65.506473 mqtt
r: 142.032718 p: -7.518594 y: -65.325637 mqtt
r: 140.789296 p: -7.402867 y: -65.775582 mqtt
r: 142.130499 p: -7.772007 y: -68.360645 mqtt
r: 142.078890 p: -7.720342 y: -74.376641 mqtt
r: 142.024959 p: -7.559341 y: -78.297491 mqtt
r: 142.393789 p: -7.536483 y: -75.006433 mqtt
r: 142.352631 p: -7.725525 y: -79.763359 mqtt
r: 142.407942 p: -7.526453 y: -77.863214 mqtt
r: 142.341825 p: -7.410883 y: -74.785736 mqtt
r: 142.305530 p: -7.338413 y: -73.173094 mqtt
r: 141.970085 p: -7.008512 y: -74.712345 mqtt
r: 142.750776 p: -7.332603 y: -82.262061 mqtt
r: 142.514452 p: -6.991238 y: -82.609396 mqtt
r: 142.260368 p: -7.231219 y: -85.826484 mqtt
r: 142.280326 p: -7.100525 y: -81.587593 mqtt
r: 142.238662 p: -7.080228 y: -80.687012 mqtt
r: 142.542797 p: -7.165315 y: -85.561534 mqtt
r: 142.571948 p: -7.250307 y: -88.150674 mqtt
r: 142.231135 p: -6.970378 y: -86.939470 mqtt
r: 142.384746 p: -7.016370 y: -90.415620 mqtt
r: 142.152656 p: -7.043447 y: -90.419683 mqtt
r: 142.175811 p: -7.032894 y: -89.562987 mqtt
r: 142.520804 p: -6.921780 y: -83.278809 mqtt
r: 142.130759 p: -6.895432 y: -82.328081 mqtt
r: 143.280252 p: -6.624256 y: -74.128105 mqtt
r: 142.847970 p: -6.889167 y: -79.196761 mqtt
r: 142.016640 p: -6.827138 y: -77.163659 mqtt
r: 142.092059 p: -6.723567 y: -79.354825 mqtt
r: 143.000187 p: -6.331878 y: -77.103704 mqtt
r: 142.642627 p: -6.389145 y: -77.435706 mqtt
r: 142.527525 p: -6.485835 y: -79.196610 mqtt
r: 142.177901 p: -6.674507 y: -75.740910 mqtt
r: 142.141455 p: -6.877830 y: -74.665285 mqtt
r: 142.218131 p: -6.653970 y: -77.679953 mqtt
r: 141.903587 p: -6.359287 y: -75.915941 mqtt
r: 142.958359 p: -6.115985 y: -71.237883 mqtt
r: 142.917952 p: -6.329769 y: -68.420416 mqtt
```

Figure 10-10 *Terminal screenshot showing the mqttFusion.py program output.*

Auto Start

The mqttFusion.py script must be configured to automatically start upon powering on the stack. It is just not reasonable to expect a caregiver to be able to, or even be interested in, using a console or ssh remote login to start the application. There are a variety of ways to auto start a Python application, but I found the easiest is to modify a standard configuration file named rc.local, which is in the etc directory. rc.local is checked every time the OS starts, and any commands found in it are always run at the root level. All you need to do to make the mqttFusion.py script auto start is to add the following line to the rc.local file:

```
python /root/RTIMULib-master/Linux/python/tests/
mqttFusion.py
```

NOTE: *Your absolute path, which is all the directories before the script name, will likely be different from what you see above. Just alter the path name to reflect where the script is located in your setup. The listing below shows my modified rc.local. It is important to place the auto start line after any other configuration commands.*

```
#!/bin/sh -e
#
# rc.local
#
# This script is executed at the end of each multiuser run level.
# Make sure that the script will "exit 0" on success or any other
# value on error.
#
# In order to enable or disable this script just change the execution
# bits.
#
# By default this script does nothing.

echo 1 >/sys/devices/virtual/misc/watchdog/disable

#/sbin/first-install.sh
python /root/RTIMULib-master/Linux/python/tests/mqttFusion.py

exit 0
```

Now that the auto start is covered, I believe that some discussion regarding the broker website would be helpful for your overall understanding of the role that the MQTT broker plays in this messaging scheme.

MQTT Brokers

The web server located at m2m.eclipse.org is a public sandbox hosted by the Eclipse Foundation as part of their open-source IoT project. This web server's software is based upon the Mosquito project created and maintained by Roger Light, a highly talented UK developer. The sandbox server allows free and public access to an actual MQTT broker where developers may test their software. There are no restrictions at this site, and just like an African waterhole, all are welcome to use it, but beware of any predators who might be lurking nearby. This metaphor means that your data being sent to the broker can be accessed by anyone who is also concurrently on the site. This is usually not a problem, as most developers are typically well behaved.

There are a number of other freely available MQTT brokers besides m2m .eclipse.org. Table 10-3 lists all the brokers that were reported as available at the time of this writing. All offer standard MQTT broker support, while some provide additional services, such as SSL, a dashboard, or an HTTP bridge, as noted in the Remarks column.

The HTTP bridge is one of the features in the m2m.eclipse.org broker that will allow us to check whether or not the mqttFusion application data is actually being received by the broker. To use the HTTP bridge, first ensure that the

Address	Ports	Add'l Services	Remarks
m2m.eclipse.org	1883, 80	HTTP bridge	Xively stats, topics, mosquito info web page
test.mosquitto.org	1883, 8883 (SSL), 8884 (SSL), 80	HTTP bridge	Xively stats, topics, mosquito info web page, SSL support
dev.rabbitmq.com	1883	Dashboard	
broker.mqttdashboard.com	1883, 8000	Dashboard	stats, HiveMQ info web page, SSL not yet available
q.m2m.io	1883		Requires registration before use
www.cloudmqtt.com	18443, 28443 (SSL)		mosquito info web page, SSL support. Requires registration; paid site but there is a free plan.

Table 10-3 *Public MQTT Brokers*

29.7

Figure 10-11 *HTTP bridge screenshot for the p subtopic.*

mqttFusion client is running, and then, using a browser on a remote computer, go to this website:

http://eclipse.mqttbridge.com/edison123/imu/p

Once on the website, you should see only a single number such as 21.4 that represents a pose coordinate reading taken from the 9DOF imu sensor. Figure 10-11 is a screenshot of the HTTP bridge website taken while I was running the mqttFusion application.

Now that you have had an introduction to the MQTT brokers, it is time to examine the MQTT clients.

MQTT Subscriber Clients

The use of a callback method is the key to how an MQTT subscriber client functions. A *callback method* is one that is triggered by an event, which, in this case, is the arrival of an Edison-published message at the broker. I will be using two third-party subscriber clients to demonstrate this functionality on a Mac and on an Android smartphone, respectively.

Mac MQTT Subscriber Client

I used a Mac MQTT client made freely available from a German website mqttfx.jfx4ee.org run by Jens Deters. Jens also has versions for Windows and Linux platforms that you can download for free. The current version, at the time of this writing, is MQTT.fx-0.0.14.3, which, I am sure, will be updated with new features and bug fixes by the time you are reading this. The MQTT.fx application was easy to install and quite intuitive to run. Figure 10-12 shows my initial configuration, which consisted of selecting the M2M Eclipse type connection from a drop-down menu and entering the full topic and all the subtopics leading to the desired data to be monitored.

It is also very easy to run this application. First, start up the stack by using the slide switch on the Battery Block, and stand the stack so that it is vertical, as

MQTT.fx - 0.0.14.3

M2M Eclipse

local mosquitto

M2M Eclipse

Connect   Disconnect

r Status   Log

edison123/imu/p   Subscribe   Unsubscribe

Figure 10-12 *MQTT.fx initial configuration.*

shown in Figure 10-13. Standing the stack this way will ensure continuous data transmission to the broker.

Wait about 45 seconds for the stack to acquire an IP address and then connect to the broker. Next start the MQTT.fx application and click on the Connect button, after ensuring that the M2M Eclipse is shown in the top-level text block. Next click on the Subscribe button, which should make the topic and all the subtopics appear in the text box just to the left of another smaller subscribe button. Finally click on the smaller subscribe button once you have confirmed that the topic and subtopics are correct. Figure 10-14 shows the results after several seconds of operation.

Remember, just seeing any data from the stack indicates that the patient is attempting to move, or is in fact getting up from the bed. This messaging scheme

Figure 10-13 *Stack positioned for continuous data transmission.*

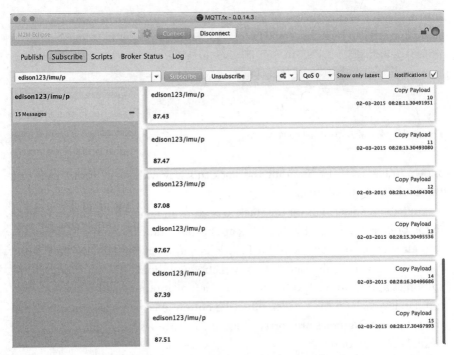

Figure 10-14 *MQTT.fx program results with data streaming from the stack.*

should provide the nearby caregiver plenty of time to go to the patient's location and check that everything is OK.

I will next demonstrate another MQTT client that uses an Android smartphone to connect to the broker.

Android Smartphone Subscriber Client

This MQTT client is named MQTT Example and was downloaded and installed from Google's Playstore application, using an Android Samsung S4 smartphone. The application is free and quite easy to use. The first time you start the application, you will see two blank text boxes at the top. Fill in these boxes with the following information:

```
m2m.eclipse.org
edison123/imu/p
```

Next click on the Connect button and you will go to another window where I had to reenter the topic and sub topic line. Next click on the Subscribe button

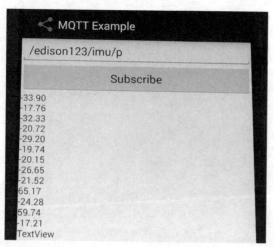

Figure 10-15 *MQTT Example application running on an Android smartphone.*

and you should start to see data streaming from the stack, provided it is situated on its end or you are shaking it. Figure 10-15 shows the results of using this Android application.

Again, any data showing on this screen indicates that the patient is up and moving.

This last section concludes this chapter on creating a useful and wearable Edison project. I hope you found it interesting and perhaps a bit useful.

Summary

In this chapter, I demonstrated a wearable Edison project. The goal of this project was to monitor an infirm patient such that the wearable would alert a caregiver if the patient was attempting to get out of the bed. I used a three-stack setup consisting of a Base Block with an Edison mounted on it, a 9DOF sensor, and a Battery Block for total portability. The stack used WiFi to connect to an M2M broker website where data could be accessed by a caregiver using an appropriate client platform. I demonstrated both laptop and smartphone applications set up to access the broker website. Any data that appeared on the website would be an indication to the caregiver that the patient was on the move.

Index